Dear Colleagues:

Manufacturing plays an outsized role in the U.S. economy: from the greatest economic multiplier of any other sector, to the creation of 4 additional jobs for every manufacturing job, it is clear that the manufacturing sector is a critical driver to our country's prosperity and security. These economic impacts grow as we add next-generation technologies: advanced manufacturing produces sophisticated and exclusive products that we can sell around the world, leading to greater economic prosperity and increasing the job multiplier to 16-to-1.

A vibrant manufacturing sector, especially one invigorated with the latest technologies, requires focused investments along the entire technology innovation pipeline. An innovation pipeline stocked with ideas, from the laboratory bench to the manufacturing shop floor, is required to "make it here" so that we can "sell it everywhere." The advanced manufacturing processes and products we desire—the types that lead to greater economic prosperity—is the result of knowledge, accumulated through research and development, translated systematically into know-how.

A vibrant manufacturing sector also needs an equally vibrant workforce, educated in a multitude of fields from engineering to economics. These skilled craftsmen, technicians, designers, planners, researchers, engineers, and managers will be in high demand: over the next decade, we will need to fill nearly 3.5 million manufacturing jobs, although 2 million of these positions may remain unfilled due to a skills gap. In fact, at this moment, 80 percent of manufacturers currently report a moderate or serious shortage of qualified applicants for skilled and highly-skilled production positions.

Growing and sustaining the innovation ecosystem we envision for advanced manufacturing will require concerted efforts across government, industry, and academia. This document contributes to a shared vision of the most important technologies for U.S. competitiveness in manufacturing by articulating a collection of advanced manufacturing technology areas and education and workforce development initiatives that are priorities for the Federal members of the Subcommittee for Advanced Manufacturing. This is a critical step towards identifying smart, strategic investments that build on our strengths—revving the engines of American ingenuity and honing the skills of the world's most productive workforce.

Sincerely,

Tom Kalil
Deputy Director for Technology and Innovation
Office of Science and Technology Policy

The National Science and Technology Council

The National Science and Technology Council (NSTC) is the principal means by which the Executive Branch coordinates science and technology policy across the Federal research and development enterprise. A primary objective of the NSTC is establishing clear national goals for Federal science and technology investments. The NSTC prepares research and development strategies that are coordinated across Federal agencies to form investment packages aimed at accomplishing multiple national goals. The work of the NSTC is organized under committees that oversee subcommittees and working groups focused on different aspects of science and technology. More information is available at: www.whitehouse.gov/administration/eop/ostp/nstc.

The Subcommittee on Advanced Manufacturing

The NSTC Subcommittee on Advanced Manufacturing (SAM) serves as a forum for information-sharing, coordination, and consensus-building among participating agencies regarding Federal policy, programs, and budget guidance for advanced manufacturing. Originally chartered in 2012, the subcommittee seeks to identify: gaps in the Federal advanced manufacturing research and development portfolio and policies, programs and policies that support technology commercialization, methods of improving business climate, and opportunities for public-private collaboration. Regarding advanced manufacturing programs conducted by the Federal Government, the subcommittee engages in the identification and integration of multi-agency technical requirements, joint program planning and coordination, and development of joint strategies or multi-agency joint solicitations.

About this Document

This document was developed by the Subcommittee on Advanced Manufacturing (SAM). The document was published by the Office of Science and Technology Policy (OSTP).

Acknowledgements

Many individuals across the Federal Government contributed to this report and the work it describes. Particular recognition goes out to the lead authors and their key subject matter experts, who crafted the input and feedback of their interagency colleagues into this single document: Bill Goldner, Sonny Ramaswamy, and Valerie Reed (USDA); Jason Boehm, Dean DeLongchamp, and Heather Evans (DOC); Kristy Pottol, Adele Ratcliff, and Tracy Frost (DoD); Mark Johnson, Reuben Sarkar, and Andrew Steigerwald (DOE); Jean Hu-Primmer and Sau L. Lee (HHS); Theresa Good and Bruce Kramer (NSF); Robbie Barbero, Megan Brewster, Austin Brown, Robert Strickling, and Lloyd Whitman (OSTP); and to Frank Pfefferkorn (AMNPO).

Copyright Information

Subcommittee Members

The following Federal departments and agencies are represented on the SAM:

Department of Agriculture

Department of Commerce

Department of Defense

Department of Education

Department of Energy

Department of Health and Human Services

Department of Homeland Security

Department of Labor

Department of Transportation

Environmental Protection Agency

National Aeronautics and Space Administration

National Science Foundation

Small Business Administration

The following offices of the Executive Office of the President are also represented on the SAM:

National Economic Council

Office of Management and Budget

Office of Science and Technology Policy

Table of Contents

Executive Summary

Advanced manufacturing drives long-term economic prosperity and growth, and supports the missions of the Federal agencies participating in the NSTC Subcommittee for Advanced Manufacturing (SAM). A foundation of priority technology areas is needed to secure U.S. competitiveness in this sector, from which collaborations between government, industry, and academia may be built. This document captures technology areas in advanced manufacturing that are current priorities for the Federal Government, and are strong candidates for focused Federal investment and public-private collaboration. Emerging technology areas include advanced materials manufacturing, engineering biology to advance biomanufacturing, biomanufacturing for regenerative medicine, advanced bioproducts manufacturing, and continuous manufacturing of pharmaceuticals. In addition, the Federal Government has already announced a number of advanced manufacturing technology areas that are either the focus of substantial existing investments or that may be the subject of future programming. These existing technology areas similarly require support across the development pipeline to fully leverage current research and development investments and infrastructure. Finally, Federal education and workforce training programs for manufacturing, which encourage strong industry involvement to ensure that today's curricula meet tomorrow's workforce needs, are highlighted.

Introduction

Advanced manufacturing[1] strengthens the U.S. economy and national security, produces high income jobs, and generates technological innovation—driving long-term economic prosperity and growth. Advanced manufacturing further supports the missions of many Federal agencies, from protecting national security and building U.S. competitiveness to strengthening the scientific and engineering enterprise and providing transformative science and technology solutions, and beyond.

Yet the importance of advanced manufacturing to winning the future was not always so clear. In fact, the U.S. manufacturing sector is still coming back from an era without this critical insight. U.S. manufacturers are currently reinvigorating plans for next-generation products that were shelved due to scarce capital, retraining workers in the newest technologies required for competitiveness in a global marketplace, and rebuilding supply chains that were dismantled by years of offshoring. The transition to advanced manufacturing processes and products is actively underway across many industries, to the benefit of the entire nation.

Sustaining this progress and securing U.S. leadership in advanced manufacturing requires participation from government, industry, academia, and other key stakeholders. This broad ecosystem must align on the key technologies that underpin U.S. competitiveness in this sector to provide a foundation of priorities for public-private collaborations and a shared vision for how to advance them. Promising technologies are those that suffer from gaps in support for the pre-competitive research and development requisite to unleashing new industries. Collaborative efforts leverage common assets to the benefit of all stakeholders, overcoming the significant development cycle costs that are impractical for any single entity to bear alone. To this end, the Federal Government may serve as an unbiased entity to convene stakeholders, to address pre-competitive technologies. Indeed, these are key findings and recommendations from the President's Council of Advisors on Science and Technology (PCAST) in its 2014 Accelerating U.S. Advanced Manufacturing report.[2]

This document captures technology areas in advanced manufacturing that are current priorities for the Federal Government and are strong candidates for focused Federal investment and public-private collaboration. Each technology area is defined, including its industry pull, cross-cutting impact, importance to national security and competitiveness, and utilization of current U.S. strengths and competencies. Major technical challenges and opportunities are outlined, and a sampling of current and planned Federal programs and initiatives is highlighted.

In addition to these emerging technology areas, the Federal Government has already announced a number of advanced manufacturing technology areas that are either the focus of substantial existing investments or that may be the subject of future programming. These existing priority areas similarly

---

[1] Advanced manufacturing Advanced manufacturing is a family of activities that (a) depend on the use and coordination of information, automation, computation, software, sensing, and networking, and/or (b) make use of cutting edge materials and emerging capabilities enabled by the physical and biological sciences, for example nanotechnology, chemistry, and biology. It involves both new ways to manufacture existing products, and the manufacture of new products emerging from new advanced technologies. President's Council of Advisors on Science and Technology. *Report to the President on Ensuring American Leadership in Advanced Manufacturing.* June 2011.
https://www.whitehouse.gov/sites/default/files/microsites/ostp/pcast-advanced-manufacturing-june2011.pdf

[2] President's Council of Advisors on Science and Technology. *Report to the President: Accelerating U.S. Advanced Manufacturing.* October 2014.
https://www.whitehouse.gov/sites/default/files/microsites/ostp/PCAST/amp20_report_final.pdf

require support across the development pipeline to fully leverage current research and development investments and infrastructure.

Scaling up advanced technologies from proofs-of-concept at the laboratory bench to world-class products on commercial shelves requires a steady stream of highly trained workers to bring the latest knowledge and skills into the workplace. A number of Federal programs captured in this document encourage strong industry involvement in manufacturing education and training, ensuring that today's curricula meet tomorrow's workforce needs.[3]

---

[3] For more information, please see http://www.manufacturing.gov/ and the following reports by the President's Council of Advisors on Science and Technology: *Report to the President: Accelerating U.S. Advanced Manufacturing* (October 2014) https://www.whitehouse.gov/sites/default/files/microsites/ostp/PCAST/amp20_report_final.pdf and *Report to the President on Capturing Domestic Competitive Advantage in Advanced Manufacturing* (July 2012) https://www.whitehouse.gov/sites/default/files/microsites/ostp/pcast_amp_steering_committee_report_final_july_17_2012.pdf, and *Report to the President on Ensuring American Leadership in Advanced Manufacturing* (link above).

Manufacturing Technology Areas of Emerging Priority

Advanced manufacturing is enabled by a multitude of technologies. In this report, the Subcommittee on Advanced Manufacturing highlights five technology areas that are focuses of widespread interest and strong support among the Federal agencies involved in advanced manufacturing. These emerging technology areas are strong candidates for future investment and expanded collaboration between government, industry, and academia.

Advanced Materials Manufacturing

Description

Whether they be lightweight automotive components made from new alloys that are stronger than steel and only a fraction of the weight, or materials that have been engineered at the nanoscale to turn waste heat into electricity, advanced materials enable the production of new products with unprecedented functions. To fully capitalize on the emergence of new advanced materials, industry requires new tools and approaches to tailor their design and quickly produce them at scale. Advanced materials manufacturing cuts across a multitude of industries—such as automotive, aircraft, biomedical, and electronics—which are pillars of our national economy and also important to our national defense. This technology area leverages the historic leadership of U.S. industry in high-technology product manufacturing, as well as its significant intellectual leadership in materials simulation and nanofabrication.

Materials matter: the efficiencies of high-temperature turbine engines, biocompatibility of replacement joints and implants, operational life of advanced batteries, and sophisticated electronics that make our digital world possible are all determined by materials that have been invented and optimized for the application. These innovations shape today's world and are essential for continued future innovation. Yet transitioning from an initial material discovery to commercial application typically takes decades. Recognizing the importance of advanced materials in supporting an innovation-driven U.S. manufacturing sector, the Administration introduced the Materials Genome Initiative[4] (MGI) in 2011 with the aim for the United States to discover, develop, manufacture, and deploy advanced materials twice as fast and at a fraction of the cost of current methods. This ambitious goal is within reach—some early successes[5,6,7] have demonstrated that a systems-level approach to material design, optimization, and implementation can significantly reduce design time and cost while improving quality. The MGI is a holistic new approach to materials science that will greatly accelerate the pace of advanced materials manufacturing technology development because the properties of materials are almost always important to the manufacturing process or product.

The advanced properties of some materials are enabled by nanoscale features. The ability to scale the production of nanoscale materials, structures, devices, and systems in a reliable and cost-effective way is a key aspect of what is referred to as nanomanufacturing. The benefits of nanotechnology stem from the ability to tailor the key structures of materials at the nanoscale to enable new or improved properties; for

---

[4] https://www.mgi.gov/

[5] *Computational Thermodynamics: The Calphad Method*, Cambridge University Press, Cambridge; B Kuehmann, B Tufts, P Trester, "Computational Design for Ultra High-Strength Alloy." *Advanced Materials and Processes* **2008**, 37; New York, **2007**;

[6] F Ronning, JL Sarrao, "Viewpoint: Materials Prediction Scores a Hit." *Physics* **2013**, 109;

[7] Horstemeyer, *Integrated Computational Materials Engineering (ICME) for Metals: Using Multiscale Modeling to Invigorate Engineering Design with Science*, Wiley-TMS, Hoboken, N.J, **2012**.

example, to make materials stronger, lighter, more durable, more reactive, or better electrical conductors. Since 2001, the National Nanotechnology Initiative[8] (NNI) has served as the vehicle for coordinating nanotechnology research activities across the Federal Government. In recent years, a Nanomanufacturing Signature Initiative[9] has brought together Federal agencies to address specific barriers to the scale-up of nanoscale materials systems, including those utilizing carbon nanomaterials in composites and nanocellulose.

Technical Challenges

The typical latency between the initial discovery of advanced materials and the ramp up from bench-scale production (grams) to full-scale, commercial manufacturing (kilograms, kilotons) is 10 to 20 years.[10] Materials design and process development must be greatly accelerated to deliver timely solutions to important national needs, such as providing clean energy, enabling next-generation electronics, and strengthening our security and national defense. Although science and engineering have always provided models for developing new materials and processes, recent breakthroughs in materials modeling, theory, high-throughput computation, and data mining can now be exploited to significantly accelerate discovery and deployment of advanced materials while decreasing their cost.

Growth in materials characterization, modeling and simulation, and data analytics is underway; these new capabilities require continued support to realize the potential of advanced materials manufacturing. Lightweight structural composites, energy storage materials, and biomedical device materials are some of the materials that are often purposefully structured down to the nanometer scale. Better methods to understand their structure will enable us to determine the origins of improved performance when a new material is discovered, and to verify that the intended structure is obtained when new materials are designed. The high-quality data collected by these methods must be stored in repositories with broad and rapid access, which will fuel the development of data analytic approaches at massive scales, providing unprecedented opportunities to discover new materials and refine or improve existing ones.

---

[8] http://www.nano.gov/

[9] http://www.nano.gov/NSINanomanufacturing

[10] The National Academies of Science, Engineering, and Medicine. *Integrated Computational Materials Engineering: A Transformational Discipline for Improved Competitiveness and National Security*. The National Academies Press. 2008.

Photo has been removed due to
copyright restrictions

Figure 1. Lawrence Livermore National Laboratory researchers fabricated a graphene aerogel lattice by three-dimensional (3D) printing into a predetermined architecture that improved its performance.[11]

Research and development in support of a basis for pre-qualified material selection is only the beginning. The material must then be incorporated into products that can be efficiently manufactured at commercial scales. Current manufacturing process designs are typically not informed by a detailed understanding of the range of process conditions a material will experience; thus, the characterization and modeling of the manufacturing process itself emerges as a key challenge, and an essential element will be the availability of high-throughput, in-line metrology to collect data throughout the manufacturing processes. From this data, models describing the process sequence will be developed, whether it be polymer extrusion, nanofabrication of electronics, nanocrystal solution growth, or the forming of sheet metal automotive parts. These models will provide design knowledge to accelerate the development of future processes and to reduce the margins needed for materials selection. To become valuable assets that enable closed-loop process control and quality assurance, in-line measurement tools must be fast, economical, and sensitive to the nanoscale.

Finally, a manufactured product must be tested to ensure it meets performance targets and safety requirements. Improved characterization methods for materials and end-user performance will be required. Because new materials may have unforeseen hazards, it is critical to develop a system that ensures that potential environmental, health, and safety issues are addressed with a strategy that is grounded in the principles of risk assessment and of product life cycle analysis.

---

[11] Zhu, C., TY-J Han, EB Duoss, AM Golobic, JD Kuntz, CM Spadaccini, MA Worsley. "Highly compressible 3D periodic graphene aerogel microlattices." *Nature Communications.* 6 (2015): 6962.

Federal Investments

The current decade(s)-long delay between material discovery and deployment in commercial products, combined with industry's increasingly shorter-term incentives to maintain cost parity, results in insufficient support for promising advanced materials manufacturing capabilities. The Federal Government must continue investments in basic research to drive innovation, while also turning the fruits of that investment into real advantages for industry, by mitigating the risk and delay inherent in transitioning to at-scale materials manufacturing through programs such as the NNI and MGI. Federal investments in advanced materials manufacturing cut across multiple agencies and impact every stage of the innovation lifecycle, from supporting cutting-edge research at U.S. universities and laboratories, to providing the infrastructure necessary to transition these advances from the laboratory to the marketplace. For example, Federal investments reported under the NNI[12] provide support for nanotechnology research and development centers and networks, including the National Science Foundation's (NSF's) Center for High-rate Nanomanufacturing focused on high-rate, high-volume nanomanufacturing tools and process. Federal agencies participating in the MGI are engaging in a number of activities[13] in support of MGI goals, including programs such as the Department of Defense's (DoD's) Automatic Flow for Materials Discovery, a U.S. Navy-funded multi-university research consortium home to a collaborative laboratory of scientists, technologists, and engineers sharing data and ideas to help equip the next generation materials workforce Other examples include the Department of Energy (DOE) Office of Energy Efficiency & Renewable Energy (EERE) Fuel Cell Technologies Office Database, which is supporting the integration of materials experimentation, computation, and theory by providing access to data that will accelerate the development of hydrogen storage materials; and the DOE Energy Materials Network for accelerated advanced materials development. Coordinated activities such as the Nanotechnology Signature Initiatives[14] provide a framework for close and targeted program-level interagency collaboration in areas including nanomanufacturing, data infrastructure, and sensor development. It is imperative that Federal agencies coordinate in these areas and look for opportunities to partner with industry and academia.

Federal agencies including DoD, National Institute of Standards and Technology (NIST), NSF, DOE, U.S. Department of Agriculture (USDA), National Aeronautics and Space Administration (NASA), and National Institutes of Health (NIH) intend to continue their support in fiscal year 2017 for advanced materials manufacturing across multiple fields. Two key goals include: (1) the development of data repositories and predictive software tools to facilitate the design of materials ranging from new structural metals to polymers that would enable directed self-assembly for new advanced electronic products; and (2) advanced sensor technology and nanofabrication tools to support the at-scale manufacture of material products including carbon-nanotube based composites, optical meta-materials, and biopharmaceuticals.

---

[12] http://www.nano.gov/centers-networks

[13] https://www.mgi.gov/activities

[14] http://www.nano.gov/signatureinitiatives

Table 1. Selected examples of Federal investment in advanced materials manufacturing.

Lead(s)	Title	Scope
DoD DARPA	Atoms to Product	This program of DoD's Defense Advanced Research Projects Agency (DARPA) seeks to develop the technologies and processes required to assemble nanometer-scale components into systems, devices or materials that are at least millimeter-scale while maintaining "atomic-scale" behaviors and characteristics. (TRL 1/MRL 1) [A]
NSF	Designing Materials to Revolutionize and Engineer our Future	This is the primary program by which NSF participates in the MGI for Global Competitiveness. This program supports activities that accelerate materials discovery and development by building the fundamental knowledge base needed to design and make materials with specific and desired functions or properties from first principles. (TRL 1-2/MRL 1-2) [B]
NSF	Materials Research Science and Engineering Centers	Materials Research Science and Engineering Centers provide sustained support of interdisciplinary materials research and education of the highest quality while addressing fundamental problems in science and engineering. These centers address research of a scope and complexity requiring the scale, synergy, and interdisciplinary interactions provided by a campus-based research center. They support materials research infrastructure in the United States, promote active collaboration between universities and other sectors, including industry and international institutions, and contribute to the development of a national network of university-based centers in materials research, education, and facilities. (TRL 1-2/MRL 1-2) [C]
NSF	Directorate for Engineering: Advanced Manufacturing	The Directorate for Engineering supports fundamental research on manufacturing and building technologies across a wide range of materials and size scales, with emphases on efficiency, economy, and minimal environmental footprint. Research is supported to develop predictive and real-time models, novel experimental methods for manufacturing and assembly of macro, micro, and nanoscale devices and systems, and advanced sensing and control techniques for manufacturing processes. The Scalable Nanomanufacturing Program supports research on new manufacturing methods to overcome the scientific and engineering barriers that prevent the production of useful nanomaterials and nanostructures at an industrially relevant scale (TRL 1-3/MRL 1-3) [D]
NASA	Physical Science Research Program	The Physical Science Research Program includes microgravity materials science research conducted on the International Space Station. NASA has recently initiated the MaterialsLab Program to execute research that will enhance our understanding of materials properties, enabling the development of higher-performing materials and processes for use both in space and on Earth. (TRL 1-3/MRL 1-3) [E]
DOC NIST	NIST MGI	In order to accelerate the discovery and development of new materials through utilization of the MGI paradigm, NIST is establishing the essential materials data and model exchange protocols and the means to ensure the quality of materials data and models, ultimately establishing new methods, metrologies, and capabilities necessary for accelerated materials development. Additionally, though its efforts to integrate these activities, NIST is working to test and disseminate its developed infrastructure and best practices to its stakeholders. (TRL 1-3/MRL 1-3) [F]
DOC NIST	Center for Hierarchical Materials Design	The Center for Hierarchical Materials Design is a center of excellence for advanced materials research focusing on developing the next generation of computational tools, databases and experimental techniques to enable the accelerated design of novel materials and their integration to industry. (TRL 1-3/MRL 1-3) [G]
DOC NIST	Material characterization for future computing systems	NIST is working to advance the measurements needed to support future electronics, including pioneering research in molecular interfaces, condensed matter physics, and two-dimensional materials such as graphene. (TRL 1-4/MRL 1-3) [H]

Lead(s)	Title	Scope
DOE EERE	Critical Materials Institute	This Energy Innovation Hub focuses on technologies that make better use of materials and eliminate the need for materials that are subject to supply disruptions. These critical materials (e.g., rare earth materials) are essential for U.S. competitiveness in clean energy. (TRL 1-7/MRL 1-7) [I]
DOC NIST	NIST Center for Automotive Lightweighting	The objective of this center is to develop the measurement methodology, standards and analysis necessary for the U.S. auto industry and base metal suppliers to transition to advanced lightweight materials for auto body components without wasteful trial-and-error development cycles, and successfully transfer this technology to customers in industry. (TRL 1-4/MRL 1-4) [J]
DOE EERE	National Laboratory-based Manufacturing Demonstration Facility	The Manufacturing Demonstration Facility (MDF) at Oak Ridge National Labs addresses critical research, development and demonstration challenges in advanced manufacturing through peer reviewed, cost-matched technical projects. The MDF focuses on additive manufacturing and carbon fiber composites. (TRL 2-4/MRL 2-4) [K]
DoD DARPA	STARnet	The Semiconductor Technology Advanced Research Network (STARnet) program's Function Accelerated nanoMaterial Engineering Center partners with leading U.S. microelectronics companies to explore and develop nonconventional nanomaterials, such as multiferroics, spintronics, and two-dimensional materials for advanced analog, memory, logic, and sensor applications. (TRL 3/MRL 3) [L]
DoD DARPA	Tailorable Feedstock and Forming	This program seeks to develop a new composite material format (i.e., feedstock) and associated processing technologies (e.g., reconfigurable forming) to reduce manufacturing complexity and enable use of advanced materials for small parts weighing less than 20 pounds at costs competitive with aluminum. (TRL 3/MRL 3) [M]
NASA	Game Changing Program	The Advanced Manufacturing Technologies project develops and matures innovative manufacturing processes and materials including: metallic joining, additive, composites, and digital manufacturing. The Nanotechnology Project is focused on the maturation, integration, and component level demonstration of high impact nanotechnologies for future use in NASA missions. (TRL 3-6/MRL 3-6) [N]
DoD DARPA	Diverse Accessible Heterogeneous Integration	The goal of this program is to create a revolutionary capability in defense electronics to realize integrated circuits by combining chiplets of gallium nitride, indium phosphide, silicon, and other device materials and technologies at the chip scale, while making the capability broadly available to designers throughout the DoD. (TRL 4/MRL 4) [O]

[A] http://www.darpa.mil/program/atoms-to-product
[B] http://www.nsf.gov/funding/pgm_summ.jsp?pims_id=505073
[C] http://www.nsf.gov/funding/pgm_summ.jsp?pims_id=5295
[D] http://www.nsf.gov/funding/pgm_summ.jsp?pims_id=503287&org=CMMI&from=home
[E] http://www.nasa.gov/mission_pages/station/research/news/materialslab
[F] http://www.nist.gov/mgi/overview.cfm
[G] http://www.nist.gov/coe/advmat/index.cfm
[H] http://nist.gov/pml/div683/grp04/nedm.cfm, http://www.nist.gov/cnst/nne.cfm, and
     http://www.nist.gov/mml/msed/functional_polymer/DimMetNanofab.cfm
[I] http://cmi.ameslab.gov/
[J] http://www.nist.gov/lightweighting/
[K] http://web.ornl.gov/sci/manufacturing/mdf/
[L] http://www.darpa.mil/program/starnet
[M] http://www.darpa.mil/program/tailorable-feedstock-and-forming
[N] http://gcd.larc.nasa.gov/projects/advanced-manufacturing-technologies and
     http://gcd.larc.nasa.gov/projects/nanotechnology
[O] http://www.darpa.mil/program/diverse-accessible-heterogeneous-integration

In support of the MGI and advanced materials manufacturing, the Energy Materials Network (EMN) aims to accelerate the materials-to-market process by integrating materials development stages from discovery through functional design and qualification, while incorporating manufacturing processing, scale-up considerations, and end-use performance. The EMN is composed of a broad set of national laboratory-led consortia for accelerated advanced materials research, development, and demonstration, where each consortium is focused on specific classes of materials and the most pressing challenges related to clean energy. These consortia will form the basis for a resource network of world-class capabilities in materials design, synthesis, characterization, manufacturing, and digital data management and informatics.

EMN Research Priorities Include:

(1) **Integration of computational and experimental research:** Increase access of computational resources to the experimental community, and vice versa, to accelerate both stages of research, development, and demonstration.

(2) **Multi-scale computational tools:** Enable the design of materials for specified lifetimes and reliabilities. Develop and experimentally validate tools that predict a material's performance at service conditions.

(3) **New methodologies and tools for modeling and validation of manufacturing processes**: Develop a suite of validated predictive tools, sensors and diagnostics, and other methodologies to qualify the ability of manufacturing processes to reliably generate the requisite materials characteristics.

(4) **Materials scale-up capabilities**: Develop equipment and automation algorithms for the manufacture of novel materials, along with experimental and computational tools to predict material properties, at industry-ready scales.

(5) **Digital data management and informatics**: Develop digital data repositories of material properties that can be networked, are machine discoverable, and are accessible to data analytics algorithms that can rapidly identify unique trends or correlations. Harmonize digital data storage standards to be extensible, accessible, reliable, and interoperable.

As part of this effort, DOE EERE will invest more than $40 million in fiscal year 2016 to establish the national laboratory-led consortia and fund competitive grant opportunities in the areas of lightweight materials, chemical reactions and catalysis, energy conversion materials, and coatings and packaging. In particular, these initial research efforts will focus on (1) low cost magnesium sheet alloys and processes compatible with automotive manufacturing; (2) next-generation catalysts for fuel cells that are free of the precious platinum group metals; (3) caloric materials that enable a paradigm shift in moving beyond today's refrigerants for use in energy efficient, environmentally friendly heat pumping technologies; and (4) new coatings, thermally superior encapsulants, and glass alternatives that enable innovative, durable, and bankable photovoltaic module form factors.

In fiscal year 2017, more than $120 million in additional funding has been requested to continue efforts and expand scope of lightweight materials and chemical reactions and catalysis. These efforts will focus on (1) low cost precursors for carbon fiber, (2) more robust catalysts with increased conversion efficiencies for more cost competitive biofuels and bioproducts, (3) novel materials systems for production of cost-competitive renewable hydrogen through advanced water-splitting, and (4) advanced low pressure hydrogen storage technologies that exceed current energy densities.

Engineering Biology to Advance Biomanufacturing

Description

Engineering biology is the design and wholesale construction of new biological parts and systems, and the re-design of existing biological systems for tailored purposes. The field integrates engineering and computer-assisted design approaches with biological research, to harness the power of biological systems to manufacture products that are of benefit to mankind; for example, the antimalarial drug artemisinin and synthetic spider silk, which may be spun into materials stronger than Kevlar. Engineering biology leverages advances in synthetic biology, along with other novel technologies that enable the predictable design of biological systems. To date, engineering biology to advance biomanufacturing has mostly focused on large-scale processes (up to millions of gallons), where the engineered cells are used to produce a fuel, chemical, protein, or biomaterial. This is different than many biomedical applications of engineered cells in fields such as regenerative medicine (as described elsewhere in this document), where modest volumes of the cells themselves may be the product (for example, autologous stem cells for tissue engineering applications). It is becoming increasingly clear that advances in engineering biology (and synthetic biology) such as genome editing could be applied broadly for the manufacture of chemicals, materials, and cells. Engineering biology as a field evolved from the existing bioprocessing expertise within the United States (which enabled the production of enzymes and protein therapeutics), and incorporates new organism engineering technologies (including synthetic biology and rapid prototyping), standardization, and interoperability. This field is well positioned to accelerate the rate of introduction of new products manufactured using engineering biology to the market and grow the U.S. bioeconomy.

The size of the U.S. bioeconomy is estimated at roughly $350 billion annually.[15] Its growth is based in part upon rapid improvements in the design-build-test-learn cycle for creating new biological products, which include advances in genetic engineering, DNA sequencing, and high-throughput prototyping of biological systems, as well as the ability to quantitatively describe the behavior of biological systems. Discoveries such as clustered regularly-interspaced short palindromic repeats (CRISPR) have led to unprecedented precision in genome editing. Federal investment over the past decade has either directly or indirectly supported many of these advance through funding of centers, foundries, small businesses and individual researchers. The United States is currently a leader in engineering biology research and its supporting science and technology fields. However, further investment is needed to translate the development of these new tools into robust technologies that will support reproducible biomanufacturing, new chemicals that will replace petrochemical derived consumer products, and new materials that will support national security. Investment by the Federal Government in biomanufacturing associated with engineering biology has lagged behind other parts of the world, especially investments that deliberately leverage industry expertise and public-private partnerships, such as the European Commission's "Framework Programmes."

---

[15] Industrialization of Biology: A Roadmap to Accelerate the Advanced Manufacturing of Chemicals (2015). Source:
http://www.nap.edu/catalog/19001/industrialization-of-biology-a-roadmap-to-accelerate-the-advanced-manufacturing

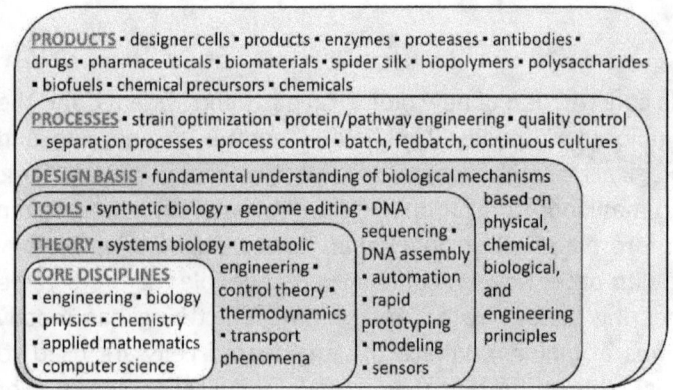

Photo has been removed due to copyright restrictions

Figure 2. (Left) Schematic of the core disciplines (white box), fundamental research (blue boxes), and applied research (tan boxes) underpinning advanced biomanufacturing. (Right). Synthetic biology can be used to engineer microbes to utilize new feedstocks, such as the sulfur shown in this artistic rendering. Federal investments in engineering biology have enabled such work. (Photo credit: Ginkgo Bioworks)

Technical Challenges

Biological systems are exquisitely complex, high-fidelity, self-replicating, adaptive, and responsive machines that utilize a wide range of starting materials and are refined by evolution. These features present intrinsic challenges and unique opportunities in engineering stable and robust biological systems for biomanufacturing purposes.

Evolution within the bioreactor is one of the inherent challenges to the design of reproducible, robust, and stable biological processes for all of biomanufacturing. Whether at the genetic or epigenetic level, all biological systems are subject to evolutionary pressures. Multiple approaches to addressing the robustness and stability of biological processes used for manufacturing will be needed. These approaches include a clearer understanding of the biological mechanisms underlying DNA stability and repair, epigenetics, and evolution, as well as development of quantitative tools that enable the prediction of complex biological designs in a manufacturing environment.

One goal of synthetic biology is the development of fully interoperable biological parts and processes—the realization of this goal, which could greatly benefit biomanufacturing, will require a deeper understanding of the complexity of biological systems, followed by an expansion of the current design space. For example, the predictable control of gene expression in eukaryotes is inherently more difficult than gene expression control in simpler organisms such as *E. coli*. Several factors contribute to challenges in expression control, such as the complexity of the organization of nucleic acids in chromatin and the highly interconnected and redundant biological circuitry (regulatory networks, metabolic networks, etc.). Also, the context of a genetic part (where the part is placed in the genome, with what spacers or insulators) dictates its performance. Further, current biomanufacturing relies largely on using existing gene-editing enzymes and chromosomal building blocks for the development of biologically derived products. There is an opportunity to expand the design space, not only in development of new enzymes that have novel reactive capacity or new proteins that have novel biomaterial properties, but also to use novel types of nucleic acids and different paradigms for codon usage. This may even open up the possibility of developing living, responsive materials. Interdisciplinary approaches that include biology, biomaterials, physics, chemistry, computer science, and engineering will be needed to address these challenges, both in the initial design and development of the new enzymes, proteins, or materials and in the translation of these processes to manufacturing.

Unlike traditional chemical manufacturing, where the environment of the chemical can be used as a reliable predictor of its probability of reaction, biological manufacturing systems (e.g., cells) are extremely sensitive and adaptable to even modest fluctuations in the environment (e.g., the reactor conditions). This makes it difficult to predict the performance of an organism in a production-scale bioreactor (potentially thousands to millions of liters) based on the performance of that same organism in laboratory-scale test tubes and shake flasks (5 to 50 mL), especially given the likely heterogeneities in heat and mass transfer at the larger scale. The fact that reactor conditions—whether in a laboratory shake flask or a thousand liter bioreactor—further influence the performance of the biological system complicates the ability to predict performance. Current approaches to address these challenges have relied on the development of rapid prototyping technologies that enable high-throughput screening of designs; these approaches can be improved upon as the failure mechanisms of the designs are fully explored. Ultimately, it may be feasible to program the organisms to function as living "chemostats" or "biostats," which sense and respond to the reactor's changing environment to self-optimize the manufacturing process and yields.

In conclusion, both a better understanding of how environment impacts organism performance (or phenotype) and better design and engineering tools for scale-up are needed to address this challenge.

Federal Investments

Federal agencies including DoD (Air Force, Army, Navy, and DARPA), DOE, Federal Bureau of Investigation (FBI), HHS NIH, NASA, NIST, NSF, and USDA National Institute of Food and Agriculture (NIFA) have invested in engineering biology for the past several years. There is significant breadth in the focus of these agencies' activities—from those focused on understanding biological complexity and enabling the development of novel technologies; to developing computational and automation tools that will facilitate the design, construction, and testing of novel engineered organisms; to a focus on agency mission-specific applications.

In addition to efforts to further the technology, NSF-funded researchers are also actively exploring how advances in engineering biology and synthetic biology will intersect with society, much in the same way as was done for the field of nanotechnology as it advanced in the early 21st century. Programs at NSF such as the Science, Technology and Society program, in collaboration with the basic science and engineering programs, currently support research in areas of public perception, risk, regulatory science, and other aspects of the technology that impact society. Other agencies have also been active in their support of research in ethical, legal, and social implications associated with engineering biology. Collaboration between agencies to date has focused on shared concerns such as public perception, environmental risk, and review of the regulatory processes associated with recent advances in the science. Workshops that focus on the environmental impact and public perception of risk associated with different technologies (including gene drives, genome editing, and possible environmental release of photosynthetic organisms) have been cosponsored and attended by representatives from many of the agencies that are investing in advances in the science.

A number of workshops convened by both NSF and NIST have been held to examine the need for standards and their community adoption. NIST has made significant progress in the industrial/commercial sector in community building around standards development and adoption.

Potential interagency collaborations to realize the commercial potential of this field include joint solicitations in areas such as: nutritional security; design of more efficient and scalable approaches to promote the bioeconomy; design of novel, high-performance materials; and development of manufacturing capabilities. Interagency efforts may further seek to leverage past Federal investments in various centers and foundries, to enable broader access to these high throughput facilities to more rapidly advance progress in the field.

Lead(s)	Title	Scope
NIH	Various programs	Production of novel technologies, therapeutics, and natural products, along with cell based therapies using engineering biology tools. NIH supports investigator-initiated projects in this area in many institutes including the National institute of General Medical Services, National Institute of Biomedical Imaging and Bioengineering, National Institute of Diabetes and Digestive and Kidney Diseases, and National Cancer Institute. (TRL 1-3/MRL 1-3) [A]
NSF	Core funding programs	Programs, including Small Business Innovation Research, in the Directorates for Engineering; Biological Sciences; Mathematical and Physical Sciences; Social, Behavioral, and Economic Sciences; and Computer and Information Science and Engineering. These core funding programs focus on basic research in the science and engineering of biotechnology, synthetic biology, including educational programs and programs focused on societal impact and risk assessment. (TRL 1-4/MRL 1-4) [B]
DoD (DARPA, Air Force Research Laboratory and Office of Scientific Research, and Army Research Office)	Various programs	DARPA's Living Foundries effort seeks to develop open-access, rapid design and prototyping foundries focused on engineering biology for the production of chemicals and materials precursors. The Air Force invests in the use of synthetic biology tools for the production of targeted metamaterials, allowing designers to create materials with defined compositions and morphology for antennas, sensors, and other applications. The Office of Naval Research seeks to biofabricate diverse products, such as inorganic materials, electronic materials, and fuels. Additional basic science efforts, notably those supported by the Army Research Office, will be useful for future manufacturing strategies. (TRL 1-3/MRL 1-3) [C]
NSF	Various Centers	These centers—including the Synthetic Biology Engineering Research Center (ERC), the Center for Biorenewable Chemicals, Science and Technology Centers, and Expeditions in Computing—focus on integrating basic research and development of complex systems as well as workforce development. Some center activities include industrial collaborations. The Synthetic Biology ERC specifically included a thrust in understanding the relationship between this new field with the public (e.g., public acceptance, legal and ethical frameworks, potential environmental impacts, regulatory burden, biosafety and biosecurity). New centers are being evaluated according to criteria including the recognition of ethical, legal, and social implications. (TRL 1-3/MRL 1-3) [D]
NIST	Foundational Measurements and Standards for Engineering Biology	NIST is working to develop and establish the tools and methods necessary to assure reliability and confidence in biological measurements impacting everything from genomic analysis and disease diagnostics to the development of cell therapies and microbial engineering. (TRL 1-5/MRL 1-5) [E]
NASA	Synthetic biological membrane and BioNutrients	This program seeks the biomanufacture of a wide range of chemicals, food, and materials that would eliminate the need to carry payload into space. Specific examples include a biomimetic lipid membrane for wastewater processing (and the ability to biologically regenerate it in-situ) and biological generation in-situ of key nutritional supplements. (TRL 2-3/MRL 2-3) [F]
DOE Bioenergy Technologies Office	Conversion R&D Program	Multiple efforts at the National Labs, as well as competitively funded projects within academia and industry. The focus is on engineering biology for the production of fuels and chemicals from lignocellulosic biomass and various waste feedstocks. (TRL 2-5) [G]
FBI	Foundry's Biosafety and Biosecurity Committee and Synthetic Yeast 2.0 Project	FBI tracks advances in engineering biology for the potential for biosecurity and dual use concerns. In support of these programs, FBI will provide insights and awareness on security topics such as dual use (intentional misuse of legitimate research/technology), illicit economies, criminal enterprise, domestic/international terrorism, exploitation, or abuse. (TRL/MRL levels not applicable) [Weblinks not available for these efforts.]

[A] http://grants.nih.gov/grants/guide/pa-files/PAR-13-137.html, http://grants.nih.gov/grants/guide/pa-files/PA-16-040.html, and http://grants.nih.gov/grants/guide/pa-files/PAR-15-285.html

[B] http://www.nsf.gov/funding/pgm_summ.jsp?pims_id=504863, http://www.nsf.gov/funding/pgm_summ.jsp?pims_id=501024, http://www.nsf.gov/funding/pgm_summ.jsp?pims_id=6673, http://www.nsf.gov/funding/pgm_summ.jsp?pims_id=503417, http://www.nsf.gov/funding/pgm_summ.jsp?pims_id=5324, and http://www.nsf.gov/eng/iip/sbir/topics/BT.jsp

[C] http://www.darpa.mil/program/living-foundries, http://www.onr.navy.mil/en/Science-Technology/Departments/Code-34/All-Programs/warfighter-protection-applications-342/Biomaterial-Bionanotechnology.aspx, and http://www.onr.navy.mil/en/Science-Technology/Departments/Code-34/All-Programs/warfighter-protection-applications-342/Synthetic%20Biology.aspx. Additional basic science efforts of note include the Institute for Collaborative Biotechnologies http://www.arl.army.mil/www/default.cfm?page=510 and http://devarajgroup.ucsd.edu/research/

[D] http://www.nsf.gov/funding/pgm_summ.jsp?pims_id=5502, http://www.nsf.gov/eng/iip/iucrc/home.jsp, and http://www.nsf.gov/od/oia/programs/stc/index.jsp

[E] http://www.nist.gov/mml/bbd/index.cfm

[F] http://gcd.larc.nasa.gov/wp-content/uploads/2015/12/FS_SSB_151217.pdf

[G] http://www.energy.gov/eere/bioenergy/processing-and-conversion

The commercial development of a new biobased renewable chemical can currently cost hundreds of millions of dollars and take more than a decade to develop. Engineering biology offers the potential to dramatically reduce the lead time and cost of bringing new renewable fuels and chemicals to market using industrially-relevant organisms, while improving the carbon conversion efficiency to desired products.

An open source Synthetic Biology Foundry[16] dedicated to engineering biology could provide the foundation for 21st century biomanufacturing by making available:

- Open strain-construction capability
- High throughput capability to test thousands of organism designs per year
- Significantly improved design-build-test-lean cycle over state-of-the-art
- Capabilities to optimize molecule production from milligrams to hundreds of grams scale
- Partnerships with many companies to identify and develop industrially-relevant host organisms
- Resources for smaller/start-up companies to speed commercialization of new molecules and organisms

Based on these needs, DOE EERE intends to establish a National Laboratory-led Synthetic Biology (SynBio) Foundry to leverage fundamental capabilities developed across multiple national laboratories to create a robust design-build-test-learn cycle and scale-up resource for rapidly, efficiently, and cost-effectively producing fuels and chemicals at scale using synthetic biology techniques. The Foundry will function as a multi-lab effort leveraging both existing and newly acquired world-class resources within the national laboratories while working with external stakeholders through competitive grant awards, cooperative agreements, and work for others; enabling partners to leverage the effort's synthetic biology tools and expertise, ultimately expediting industrial adoption of the technology.

The SynBio Foundry will establish a robust set of biomanufacturing principles, which would develop and use standardized DNA elements and commercially relevant and optimized host organisms going beyond current yeast and E. coli paradigms. From this foundation, the SynBio Foundry will seek to develop improved systems capability, processes for predictable scale-up, machine learning and mechanistic models, and standards and interoperability protocols. The SynBio Foundry will deliver a set of design-build-test-learn cycle tools and an organism development package that will be easily transferred to the biotechnology industry, enabling the scale-up of multiple, high-impact chemicals in multiple industrially-relevant host organisms.

The development of a SynBio Foundry will build on $3 million in seed funding beginning in fiscal year 2016 with an additional requested $35 million in fiscal year 2017. The requested funding will leverage recently developed synthetic biology resources to improve efficiencies in the conversion of biomass to fuels and related products. Funding may be further enhanced by other agencies that choose to fund activities within the SynBio Foundry.

---

[16] The term "foundry" is used to emphasize that a standardized set of tools, parts, and chassis organisms (e.g., analogous to a set of production molds) can be combined with a set of interchangeable genomic parts to churn out, at very high-throughput, a large variety of different fuel and chemical products. Reusing the base set of chassis for each new product minimizes the amount of the process that needs to be designed from scratch. The term "foundry" as used here has a distinctly different meaning than a purpose-built facility such as a steel foundry.

Biomanufacturing for Regenerative Medicine

Description

Regenerative medicine, and the clinical use of stem cells, has the potential to repair or replace dysfunctional, degenerating, or absent cells, tissues, and organs. Such developments may one day restore the form, function, and appearance to our severely injured service members, dramatically reduce waitlists for organ transplants, increase the availability of essential cell-based therapies, and possibly reduce healthcare cost for treatments. Additionally, engineered cells can be used to redirect immune function and enable the emergence of "immuno-oncology." Microphysiological systems ("organ-on-a-chip"), a small living and working model of a specific tissue or organ type, can greatly accelerate screening of drug candidates, probe disease mechanisms, and explore novel therapies. To realize the full potential of regenerative medicine, active cells, tissues, and organs (such as cardiovascular, renal, and neurologic) must be bioengineered and manufactured at scale.

Photo has been removed due to
copyright restrictions

Figure 3. Manufacturing of a bioengineered vascular conduit. (Left) A human acellular vessel. (Right) Phase 2 manufacturing system for bioengineered, human acellular vessel. (Photo credit: Humacyte)

The Federal Government invested $2.89 billion between 2012-2014 in regenerative medicine, primarily for the development of therapeutic techniques and technologies.[17] While there has been a steady increase in funding to demonstrate the possibility of regenerative biologic solutions to some vexing health concerns, there is a critical and growing need to focus on process engineering to achieve manufacturing reproducibility to increase the reach of the emerging therapeutics.

Technical Challenges

A significant challenge for development of regenerative medicine products and therapies is that the field is still transitioning from an academic enterprise to an industry-based, commercial enterprise. This transition requires a rigorous focus on starting materials, process, and final product characteristics to assure that products are safe, effective, and have a known potency and predictable shelf-life. The continued maturation of the regenerative medicine field is dependent on designing and developing robust production-scale manufacturing techniques, a refined set of measurements and standards that provide confidence in the product, evolutionary governance, harmonized regulations, workforce development, and supply chain diversification to support responsible and rapid development of regenerative medicine (cell, tissue, and organ bioengineering) technologies.

Important progress has been made in the basics of cell-based therapy and diagnostics, including: obtaining

---

[17] U.S. Government Accountability Office. *Regenerative Medicine: Federal Investment, Information Sharing, and Challenges in an Evolving Field*. July 23, 2015. http://gao.gov/products/GAO-15-553.

high-quality starting cells under current good manufacturing practices, improving the understanding and ability to control cellular and tissue phenotype (particularly for stem cells), developing bench-level systems for making therapies from cells and/or cells within an extracellular matrix, and producing large banks of pluripotent stem cells.

A fundamental challenge is the development of appropriate growth media for cells, in order to grow cells at the necessary scale for practical therapies and to assure consistency of product. Current systems are too inefficient for suspension-grown cells, largely non-existent for adherent cells, and patient-specific for engineered tissue. Research on efficient, safe production of cells and engineered tissues at commercial scale is required to ensure that these precision therapies will be economically and technically achievable.

The regenerative medicine therapeutics industry can learn from the traditional biopharmaceutical industry, in terms of how to control manufacturing products and processes. For example, regenerative medicine and pharmaceuticals share some conceptual challenges including bioreactor control, continuous manufacturing, scale-up of cell culture, distributed manufacturing of small personalized batch therapies, starting materials, and the use of disposables in manufacturing. Further, a goal of cell-based bioengineering is to strive towards the process control methods currently in development for pharmaceutical continuous manufacturing (as described elsewhere in this document). Although each cell and tissue product is fundamentally unique, their bespoke manufacturing processes may follow a common process governance structure where manufacturing activity is quantified, standardized, and monitored.

Manufacturing cell and tissue products is significantly more challenging than manufacturing biopharmaceuticals. This is because manufacturing release criteria for bioengineered cells, tissue, and organs are largely undefined; the mechanisms by which these living products produce their effects are poorly understood; and they contain active cells which continue to grow and evolve. Additional challenges for manufacturing cell-based products stem from the complexity of the product and the bioreactor process. These challenges include appropriate design of bioreactors for adherent cells; characterization of the product with measurements that predict the clinical effectiveness of the product; determining tolerances of variation in cell types; how to store, ship, and deliver product with a very short (hours-long) shelf life; and how to assess the purity of the product in minutes or hours, instead of days. Development of methods to address the long-term storage of cells as manufacturing inventory is essential to increase their flexibility, reliability, and cost-effectiveness.

As a result of the inherent complexity of these products, even with the application of advanced measurement science, effective manufacturing will continue to rely to some degree on the spirit of "the process is the product." This time-tested axiom of biologics[18] manufacturing emphasizes the importance of managing the local environment at every step, as well as the unique challenges for each cell or tissue type, particularly when the therapeutic interest relies on active living cells over time. Current processes are poorly controlled, inefficient, and often inadequate for growing cells in suspension culture or on micro-carriers, while large-scale processes to grow cells within their extracellular matrix do not yet exist.

---

[18] Biologics can be composed of sugars, proteins, or nucleic acids or complex combinations of these substances, or may be living entities such as cells and tissues.

The final commercial manufacturing process for any product is the purview and responsibility of the developer; however, Federal investments in development of generalizable approaches to common manufacturing design, development, scale-up, and scale-out challenges could serve to de-risk development of numerous products in the field. This may make the technology attractive enough for the pharmaceutical industry to adopt and use to develop a new class of therapeutics. One deliverable within five years could be a prototype for cell production on a larger scale: scaled-up, closed, and fully automated platforms with in-line environmental monitoring, continuous assays for cell phenotype/potency, cost/material efficiencies, and separations/purification systems for cells grown in suspension culture (e.g., red and white blood cells). Over ten years, adherent cells for organs (e.g., heart, liver, gut, lung) are desired, ideally within their appropriate matrices.

Ongoing programmatic funding continues from DoD, Food and Drug Administration (FDA), NIH, NIST, DOE National Nuclear Security Administration, NSF, and the Department of Veterans Affairs (VA) in developing the technologies that make up the regenerative medicine field.

Future support of regenerative medicine manufacturing technologies will continue to exhibit strong multidisciplinary and collaborative approaches. For example:

- To manufacture specific tissues, a project team must have advanced engineering skills combined with a deep understanding of the physiology of those tissues and the organ systems in which those tissues function.
- Statistics and computational modeling will be critical for designing and testing the platforms that will shape the product development plan and the regulatory strategy.
- Controlling and characterizing the biologic activity through discovery, triggers, and terminations of the appropriate effector and control mechanisms at the individual cell and population levels will require the expertise of genomic scientists.
- A fundamental knowledge and understanding of developmental biology is required to deliver an adult phenotype and to regulate growth reproducibly over time.
- Efficiencies at each step are essential to keep costs and raw materials usage down, and to keep fragile products (i.e., cells, tissues) viable, while a constant focus on usability will ultimately help enable adoptability.

These technical challenges, among others, must be solved to advance this field into common medical use. Advancements in research and development (R&D) will be most efficiently translated to commercial products through close collaboration with the private sector.

Table 3. Selected examples of Federal investment in biomanufacturing for regenerative medicine.

Lead(s)	Title	Scope
NSF CBET	Early Concept Grants for Exploratory Research	NSF's Chemical, Bioengineering, Environmental and Transport Systems (CBET) division uses these exploratory research grants for Cellular Biomanufacturing to investigate key challenges in scale-up, such as efficiency of manufacture, reproducibility, and viability. (TRL 1-3/MRL 1-3) [A]
DOC NIST	Advanced Manufacturing Technology	The Cell Manufacturing Consortium is currently funded to establish and strength industry-focused research consortia in order to identify future research directions and needs and create a roadmap for cell-based manufacturing which integrates research with the development of commercially viable manufacturing processes. (TRL 1-5/MRL 1-5) [B]
DOC NIST	Measurements and Standards for Regenerative Medicine	NIST is developing fundamental measurement technologies (such as new image analysis technologies for cell characterization), collaborating with industry on immediate measurement needs (such as cell counting and viability), and developing international standards with industry and FDA that are necessary to enable the development of and the quality assurance of regenerative medicine products. (TRL 1-5/MRL 1-5) [C]
HHS FDA	Mesenchymal Stem Cell Consortium	The Mesenchymal Stem Cell Consortium aims to develop strategies to facilitate the development and translation of stem cell-based regenerative medicine products. Because commonly used measures of cell identity and biological activity are often not well-correlated with and predictive of intended clinical outcomes, the consortium uses Mesenchymal Stem Cells as a prototype cell to identify more predictive measures to characterize cellular therapies. Such analytical techniques provide a potential approach to address variability issues presented by cell source and/or manufacturing conditions. (TRL 2-4/MRL 2-4) [D]
DoD (Army-led, Navy, Air Force), VA, NIH, and DHA	AFIRM	The Armed Forces Institute of Regenerative Medicine (AFIRM) seeks to develop advanced restorative, diagnostic, and cell and tissue-based treatment options for our severely wounded service men and women. (TRL 2-7/MRL 1-6) [E]
DoD DARPA, HHS NIH, and FDA	Various Microphysiological Systems programs	These programs are taking diverse approaches to develop microphysiological ("organ-on-a-chip") systems and determine the feasibility of an *in vitro* platform that uses human tissues to evaluate efficacy, safety, and toxicity of promising therapies. DARPA (TRL 2/MRL 2), HHS NIH and FDA (TRL 4/MRL 3) [F]
HHS NIH NHLBI	Stem Cell-Derived Blood Products for Therapeutic Use Program	This program aims to address the scientific questions that remain despite considerable progress towards enabling and accelerating the use of stem cell-derived blood products as therapeutics. It also supports small business research to develop improved techniques and tools to enhance the production of clinically-relevant, functional stem cell-derived red blood cells or platelets in a more efficient and cost-effective manner. (TRL 4/MRL 3) [G]
HHS NIH NHLBI	Production Assistance for Cellular Therapies III	This is the continuation of a program to advance cellular therapy research in the areas of regeneration of damaged/diseased tissues, organs, biologic systems, and targeted treatments for serious diseases without effective therapies. This program provides qualified investigators with consulting, manufacturing, preclinical study, and regulatory expertise necessary for the development of novel cellular therapies in the area of heart, lung, and blood cellular therapy. (TRL 5/MRL 5) [H]

[A] http://nsf.gov/pubs/2015/nsf15065/nsf15065.jsp

[B] http://www.nist.gov/amo/amtech/index.cfm

[C] http://www.nist.gov/mml/bbd/cell-therapies.cfm

[D] http://www.fda.gov/BiologicsBloodVaccines/ScienceResearch/BiologicsResearchAreas/ucm127182.htm

[E] http://www.afirm.mil

[F] http://www.darpa.mil/program/microphysiological-systems **and** http://www.ncats.nih.gov/tissuechip

[G] http://grants.nih.gov/grants/guide/rfa-files/RFA-HL-15-030.html

[H] http://www.fbo.gov/spg/HHS/NIH/NHLBI/NHLBI-ECB-HB-2016-09-JB/listing.html **and**
http://www.fbo.gov/spg/HHS/NIH/NHLBI/NHLBI-ECB-HB-2016-08-JB/listing.html

Bioreactor technology, which supports biologically-active environments, is an essential part of the research and development pathway. Recent trends have created a large potential market for breakthrough medical solutions led by tissue-engineered medical products. Yet, research successes for a variety of simple bioreactor systems have shown sustained, but limited, *in vivo* function. The NIH has an ongoing program to support multidisciplinary small business teams in the development of complex, three-dimensional engineering systems for growing heart, lung, or bone marrow tissue.[19] Integrated devices will require a diverse array of scientific principles and technologies. Ultimately, bioreactor designs should provide the most physiologically relevant environment to promote correct three-dimensional tissue growth and maintenance, which is also efficient, safe, and economical. Such devices should be made commercially available and be widely disseminated to researchers for application in translational science. These devices could support, for example, bioengineered trachea, vascular grafts, and other hollow-organ tissues that have made entry into clinical studies. Sophisticated and more standardized bioreactors will be critical for developing these and other complex, neo-organ technologies.

Bioreactor systems that deliver standardized, robust, and reliable lots of engineered tissue at scale for clinical and/or industrial use are not widely available. At a minimum, these platforms will need continuous monitoring (with high-content, functional assays) along the entire production line, automation to increase predictability and speed while minimizing possibilities for contamination, and supply chains for raw materials that are reliable and affordable. Additional needs and opportunities to grow and mature precursor cells at scale include devices for the growth of stem cells, which should integrate stem cell expansion with subsequent mechanical stimulation to enhance functional differentiation for use in complex tissues. Some tissues may require electrical pacing and conditioning as well. Another challenge for some cellular propagation is to establish stable protocols that do not require feeder layers and conditioned medium. At the organ level, tailored lung tissue bioreactor design plans are needed for gas and blood exchange interfaces, whereas design of cardiac tissue bioreactors might emphasize electrical and mechanical forces. Additionally, it may be possible to engineer three-dimensional bone marrow organs that permit extensive blood stem cell self-renewal and hematopoiesis *in vitro*. Given the complexity of engineered constructs—often containing a combination of cells, scaffolds, and other factors—special challenges exist for tissue engineered product characterization, growth, and manufacturing protocol.

---

[19] http://grants.nih.gov/grants/guide/rfa-files/RFA-HL-15-004.html

Advanced Bioproducts Manufacturing

Description

The United States currently relies on the average use of more than 19 million barrels per day of petroleum for fuels and as a feedstock to make products ranging from chemicals to plastics to many everyday items. Bioproducts—high-value chemicals, bioreagents, materials, fuels, and other biobased intermediates derived from renewable biological resources such as agricultural and forest waste—hold strong promise to reduce this petroleum use and serve as the backbone of an emerging bioeconomy.

The United States currently uses about 400 million dry tons of biomass annually for fuels, heat, materials, and industrial enzymes with significant environmental and economic benefits.[20] U.S. revenues from industrial biotechnology (fuels, enzymes, and materials) reached at least $125 billion in 2012 with 1.5 million direct jobs across all sectors of the supply chain.[21] This economic impact has the potential to further grow the bioeconomy by an additional $100 billion by 2030, with the creation of an additional one million jobs, particularly in rural areas where feedstocks are grown.[22] Bioproducts will reduce greenhouse gas emissions and provide a stable domestic source for these products.

While the United States has made great strides in promoting the use of sustainably-produced feedstocks to fuel economic activity and growth, the bioeconomy is still in its early stages. Significant work remains to increase the use of bioproducts to replace a variety of petroleum-based fuels and products, make those bioproducts more cost-effective relative to petroleum-based products, and further improve the sustainability and environmental benefits of bioproducts.

---

[20] https://www.whitehouse.gov/sites/default/files/microsites/ostp/national_bioeconomy_blueprint_april_2012.pdf

[21] http://www.biopreferred.gov/BPResources/files/EconomicReport_6_12_2015.pdf

[22] http://www.biomassboard.gov/pdfs/farb_2_18_16.pdf

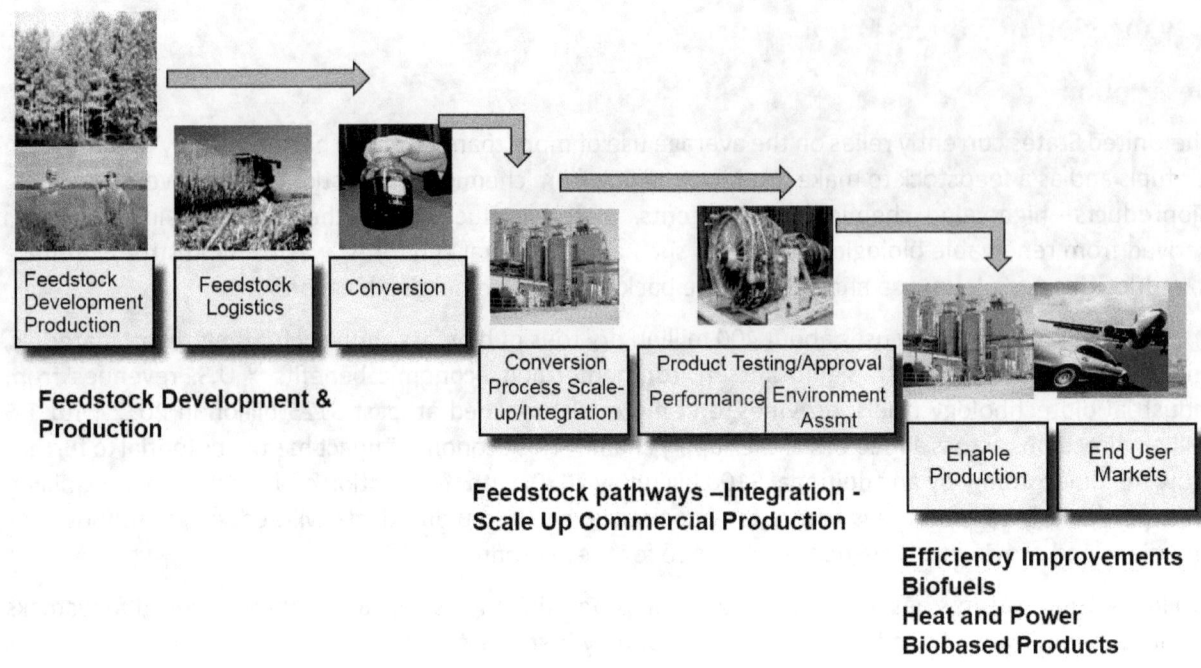

Feedstock Development & Production

Feedstock pathways –Integration - Scale Up Commercial Production

Efficiency Improvements
Biofuels
Heat and Power
Biobased Products

Figure 4. A supply chain approach to advanced bioproducts manufacturing.[23] (Photo credit: DOE BETO)

Technical Challenges

The *U.S. Billion-Ton Update*[24] reports that the United States has the resources to produce significant volumes of bioproducts and biofuels, potentially replacing up to 30 percent of current petroleum use with renewable biomass in the next 15-20 years. Under those projections, the bioeconomy could provide feedstocks for the production of biofuels required by the Renewable Fuel Standard[25] and beyond. For the United States to realize the full potential of a domestic bioeconomy, key barriers need to be addressed across the entire supply chain, including feedstocks, conversion, and integration.[26] Each of these areas face significant manufacturing challenges, as described here.

To provide a sustainable, reliable, and cost-effective supply of feedstocks appropriate for a variety of end uses, existing barriers such as cost, land use, and regional and source variations must be overcome. The inherently diverse nature of biomass feedstocks requires a fully-integrated logistics supply chain that includes harvesting practices, biomass preprocessing, transport, and storage systems tailored to the feedstock type. Today, the existing biofuels industry primarily employs conventional logistics systems developed for traditional agriculture and forestry systems. These are designed to move biomass short distances for limited-time storage (less than one year). Expanding usable feedstocks to include crop residues and purpose-grown energy crops can provide the raw materials for the manufacture of future bioproducts with superior environmental benefits. Conventional logistics systems are not well configured for a diverse, much larger set of feedstocks that are spread over larger areas. This currently creates unique transportation requirements, especially in areas with medium to low yields. Advanced logistics systems designed to deliver feedstocks with predictable physical and chemical characteristics, longer-term stability

---

[23] http://energy.gov/sites/prod/files/2015/04/f22/mypp_beto_march2015.pdf

[24] http://energy.gov/eere/bioenergy/downloads/us-billion-ton-update-biomass-supply-bioenergy-and-bioproducts-industry

[25] http://www.epa.gov/renewable-fuels-standards-program

[26] http://energy.gov/qtr

during storage, and high-capacity bulk material handling characteristics can facilitate economically viable transport over longer distances and ultimately lower the costs of delivery. Additionally, each conversion technology places specific demands on the feedstock quality and quantity that must be delivered to the biorefinery.

The blending of feedstocks offers a possible solution to quality, cost, and regional diversity issues. Feedstock blending allows a biorefinery to collect less of any one feedstock, enabling biorefineries to pay a lower average price while reducing supply risk. An improved blend of feedstocks can meet refinery requirements such as lignin content while reducing the impact of potential supply disruptions, ultimately making more feedstock available from and across regions. Feedstock blending also hedges against unintended supply disruptions to any one blend component. Preliminary results from two lignocellulosic biomass conversion facilities suggest that blending multiple preprocessed feedstocks enables the acquisition of higher biomass volumes and reduces feedstock variability to meet biorefinery in-feed specifications, while delivering feedstock to the biorefinery at $80/dry metric tonne.[27]

Conversion is the overall process of converting delivered feedstocks into the variety of final fuels and products at the biorefinery; efficient, low-cost, reliable conversion technologies are a key requirement for scaling up advanced bioproduct manufacturing. Efficiently introducing feedstocks into a pressurized reactor at large scale is problematic due to feedstock inconsistencies, such as varying properties of moisture, ash, and contaminants. In addition to challenges of introducing dry feedstocks, new reactor processes and designs are needed for the feeding of wet feedstocks, such as waste cellulosic and wet algal biomass slurries into reactors. The preprocessing of these feedstocks to remove contaminants and ash must be tailored to the needs of the conversion catalysts and demonstrated at large enough scale to indicate commercial viability.

Conversion technologies range from low-temperature biological catalysis (including organisms) to high-temperature, high-pressure chemical catalysis. The development of a single conversion technology from TRL 5 to TRL 8 can take ten years in a supportive policy environment.[28] To accelerate this timeline, dedicated investment in catalyst development is necessary. For example, a growing number of researchers are developing the tools and fundamental understanding required to engineer biological systems (as described elsewhere in this document). Continued expansion of the bioeconomy requires reduction in the time and cost associated with the design-build-test-learn cycle and ultimate scale-up of conversion technologies.

Finally, feedstock and process variations can cause fouling, plugging, corrosion, or other disruptions in biorefinery operations. The lack of operational data on fully integrated systems over extended periods of time presents large scale-up risks, thus representing a major barrier to a successful biomanufacturing industry. An improved understanding of process integration is essential for: (1) characterizing the complex interactions that exist between unit operations, (2) identifying impacts of inhibitors and fouling agents on catalytic and processing systems over commercial operation timeframes, and (3) enabling the generation of predictive engineering models that can guide process optimization or scale-up efforts and enable process control.

Federal Investments

A robust and stable supply chain connecting diverse feedstocks to production facilities and markets is

---

[27] From the Sugar Platform to Biofuels and Biochemicals, Final Report for the European Commission Directorate, General Energy, April 2015

[28] From the Sugar Platform to Biofuels and Biochemicals, Final Report for the European Commission Directorate, General Energy, April 2015

essential to the success of advanced bioproduct manufacturing at scale. Supported by public-private collaborations, the Federal Government is continuing its systematic effort to expand the sustainable production and use of biomass by supporting research and development activities across the entire supply chain. The broad array of Federal activities is coordinated through the Biomass Research and Development Board co-chaired by the DOE and USDA.[29]

Initial Federal investments have helped establish some of the requisite processes and platforms needed to generate high-value intermediate and end-use products, supporting the growth of bioproduct manufacturing in the United States. Examples include genomic research on bioenergy feedstock crops and unique processes and microbes to convert feedstocks into bioproducts, exploration of sustainable management practices, development of biomass conversion processes and expansion of bioenergy infrastructure, and cost/benefit estimates of renewable energy and byproducts production. These investments leverage corporate research and private investment, leading to more applications, lower costs, and superior sustainability of future renewable fuels and chemicals.

Additional public-private investment focused on cross-cutting bioproducts manufacturing capabilities is needed, for example to:

- demonstrate commercial-scale biomass feedstock delivery systems taking advantage of the regional feedstock supplies of wastes, residues and energy crops;

- establish common processes and chemical platforms leading to the efficient yield of high-value intermediates that can be used in the production of a diverse array of end-use bio-products (for example, hydrolysis of mixed biomass streams to sugar intermediates for further downstream processing to fuels and chemicals); and

- support commercialization through risk reduction and scale up of common process steps (such as feedstock pre-processing and handling) that will benefit a wide array of processing configurations to meet the needs of the diverse feedstocks and end products.

Such multi-stakeholder investment will serve to increase U.S. leadership in the innovation of agricultural bioproducts manufacturing for industrial adaptation and commercial applications and grow new agroindustry.

---

[29] http://www.biomassboard.gov/

Table 4. Selected examples of Federal investment in advanced bioproducts manufacturing.

Lead(s)	Title	Scope
NSF Chemical, Bioengineering, Environmental and Transport Systems	Energy for Sustainability Program	Fundamental research on innovative approaches that lead to the intensification of biofuel and bioenergy processes is an emphasis area of this program. Includes biological, thermochemical, or thermocatalytic routes for the conversion of lignocellulosic biomass to advanced biofuels beyond cellulosic ethanol, microbial fuel cells and production of hydrogen, hydrocarbons and lipids from autotropic, phototrophic or heterotrophic microorganisms. (TRL 1-3/MRL 1-3) [A]
DOE Bioenergy Technologies Office	Feedstock Logistics and Advanced Algal System Program	This program supports intramural and extramural applied R&D to develop new feedstock logistics technologies and landscape design systems that reduce costs along each step of the bioenergy supply chain including algal production systems. (TRL 2-5/MRL 2-5) [B]
DOE Bioenergy Technologies Office	Conversion R&D Program	This program supports intramural and extramural R&D on conversion pathways using varied feedstocks and technologies for the production of higher value bioproducts, in support of increased production of biofuels in a biorefinery. Work includes both cellulosic and algal-derived bioproducts, to improve the overall economics of fuels, and development of downstream processing, to upgrade biomass derived intermediates into fuels and bioproducts. (TRL 2-5/MRL 2-5) [C]
USDA NIFA	Agriculture and Food Research Initiative Programs	NIFA supports extramural research, development, and demonstration of the whole supply chain for the sustainable production of bioenergy and bioproducts, including feedstock production and conversion processes and nanocellulosic materials and products. Efforts also include technical, economic and life-cycle analysis, and education and workforce development to support the growing bioeconomy. (TRL 2-7/MRL 2-7) [D]
USDA NIFA, Agriculture Research Service, and Forest Service	Various programs	Intramural and extramural basic and applied research in conversion (fuels, chemicals, products). (TRL 2-6/MRL 2-6) [E]
DOE Bioenergy Technologies Office	Demonstration and Market Transformation	These extramural pilot- and demonstration-scale integrated biorefinery projects are cost-shared with industry to de-risk first-of-a-kind technologies and prove out integrated processes at scale. Acceptable biorefinery demonstrations include production of bioproducts in support of the overall economics of the biorefinery. (TRL 6-8/MRL 6-8) [F]
USDA Rural Development	Biorefinery Commercialization Assistance Program	This program provides commercialization assistance (loan guarantee) to first-of-a-kind biorefineries. (TRL 8-9/MRL 8-9) [G]
DoD Navy/ Defense Production Act	Advanced Drop-in Biofuel Production Project	Commercial Scale alternative fuel production in integrated biorefineries, cost shared with DOE and USDA. (TRL 8-9/MRL 8-9) [Weblinks not available for these efforts.]

[A] http://www.nsf.gov/funding/pgm_summ.jsp?pims_id=501026&org=CBET&from=home

[B] http://www.energy.gov/eere/bioenergy/biomass-feedstocks and http://energy.gov/eere/bioenergy/algal-biofuels

[C] http://www.energy.gov/eere/bioenergy/processing-and-conversion

[D] http://nifa.usda.gov/program/agriculture-and-food-research-initiative-afri

[E] http://nifa.usda.gov/funding-opportunity/biomass-research-and-development-initiative-brdi, http://www.ars.usda.gov/main/main.htm, and http://www.fs.fed.us

[F] http://www.energy.gov/eere/bioenergy/integrated-biorefineries

[G] www.rd.usda.gov

Integrated Biorefinery Demonstration at Semi-Works Scale

The pulp and paper industry is the largest producer and user of biomass energy in the United States, representing the ideal platform to launch the emerging integrated biofuel and bioproduct industries. In 2012, the Joint USDA-DOE Biomass Research and Development Initiative awarded $7 million, cost-shared with industry, to enable a demonstration of an integrated biorefinery at the semi-works scale, while allowing the participating company to increase its revenue streams and preserve 121 jobs. Domtar, a Fortune 500 company, is the largest manufacturer of uncoated freesheet paper in North America, processing well over 20 million green tons of biomass per year. Building on work initiated by Domtar in 2010, the initiative funded a three-year project to construct a semi-works demonstration plant. This involved the demonstration at-scale of two main technologies, acid hydrolysis of biomass and lignin removal technology, to convert low-value mill side streams and waste streams into higher-value sugar, tall oil, and lignin intermediates for drop-in markets. (The logistics for feedstock for wood-based biorefineries are already commercially available through an established supply chain and is nearly all derived from sustainably-grown trees.) Ultimately, Domtar successfully installed a commercial-scale lignin separation plant in Plymouth, North Carolina.[30]

---

[30] http://www.reeis.usda.gov/web/crisprojectpages/0225162-integrated-biorefinery-at-the-domtar-plymouth-north-carolina-pulp-mill.html

Continuous Manufacturing of Pharmaceuticals

Description

Continuous manufacturing is the integration of multiple manufacturing process systems into a single system, based on model controls, to enable continuous product flow and recovery as input raw materials are added to the manufacturing process. Pilot studies in the pharmaceutical and biotechnology industries suggest that continuous manufacturing may have a multitude of benefits in these industries, such as: reducing the manufacturing facility footprint by 10 to 100 times; eliminating intermediate product batches and their associated storage and testing; reducing the amount of incoming raw materials and final product waste; streamlining manufacturing processes and shortening manufacturing cycle times; increasing production yields and overall product manufacturing efficiency; improving product quality with advanced control systems; and enabling real-time release testing. Continuous manufacturing may reduce manufacturing costs, which currently consume as much as 27 percent of the revenue for many pharmaceutical companies, by up to 40 to 50 percent.

Continuous manufacturing strategies are widely used in many private sectors including food, chemical, and petroleum industries. However, the broad adoption of such strategies has lagged behind in the pharmaceutical and biotechnology industry because of the product development paradigm. This paradigm does not emphasize process optimization due to the uncertainty concerning the clinical outcome and the historical regulatory environment that drove production towards fixed "lock-in" processes for the entirety of the product lifecycle. The broad industry interest in continuous manufacturing includes small-molecule and biologic manufacturers; contract manufacturing organizations; and equipment, instrument, control, and excipient supply companies. The implementation of end-to-end continuous manufacturing (i.e., the integration of drug substance and drug product manufacturing into a single continuous process) requires leveraging U.S. capabilities in pharmaceutical sciences, systems engineering, and real–time sensing.

Continuous manufacturing promises to improve the agility, flexibility, and robustness in the manufacture of pharmaceuticals, helping to mitigate the public health threat arising from drug shortages. The advent of personalized medicine, as well as drugs seeking orphan drug designation, necessitates the production of smaller batches of drugs due to the smaller target-patient populations. A similar concept applies to medical countermeasures development for low probability-high consequence events—such as those caused by chemical, biological, radiological/nuclear, and emerging infectious disease threats—as these medical countermeasures likely do not have high commercial promise except to the Federal Government for national preparedness purposes. The many benefits of continuous manufacturing (e.g., through flexible platform technologies that utilize modular or plug-and-play equipment) may enable the rapid, domestic production of medical countermeasures, enhancing U.S. preparedness and response.

Photo has been removed due to copyright restrictions

Figure 5. Schematic diagram of continuous manufacturing, including a photograph of a first-of-its-kind continuous finished dosage form process for a recently approved cystic fibrosis drug.[31] The footprint of this tableting facility is approximately 4 percent of that of the footprint of a traditional facility needed to do the same process. (Photo credit: Vertex)

Technical Challenges

Challenges to continuous manufacturing adoption include both process development technical hurdles and scientific and regulatory challenges involved with navigating away from a "batch-centric" regulatory paradigm.

The development of a continuous manufacturing design requires a systems-based approach and the integration of multiple existing technologies. Advances in process modeling and simulation may facilitate the optimization of equipment configurations, manufacturing routes, and advanced control systems; as well as connecting the impact of material properties with model parameters to generate models that are transferrable from one formulation to the next. Modular or plug-and-play type equipment with re-usable, flexible, or interchangeable parts that can be connected in a different order are needed to enable flexible manufacturing and platform technologies for continuous manufacturing. This is especially important for certain specialty (orphan or breakthrough) drugs in the pharmaceutical industry, which may be manufactured using equipment that can be configured to produce small quantities of a diverse portfolio of products.

The manufacture of a traditional small molecule drug substance usually requires many novel chemical synthesis steps, necessitating multiple rounds of isolation and purification, adding complexity to the design of a flexible integrated continuous process. The switch to continuous processing may necessitate new process technologies and synthetic routes to streamline the system. To accomplish this, individual

[31] http://www.fda.gov/NewsEvents/Newsroom/PressAnnouncements/ucm453565.htm and http://connect.dcat.org/blogs/patricia-van-arnum/2015/09/18/manufacturing-trends-in-continuous-mode

drug manufacturers currently require expensive, highly specialized equipment that may not be commercially available; these bespoke and typically proprietary investments generate information asymmetries in the field, hindering the development of platform technologies. Development of flow chemistry toolboxes, including highly selective chemistries that allow the use of simple and effective continuous purification technologies, is needed.

The manufacture of larger molecule biotechnology products utilizes biological substrates to produce the final product (e.g., perfusion and fermentation systems). The established configuration of perfusion bioreactor-cell retention devices will likely remain conceptually similar in the future, but improvements are necessary to realize a truly continuous platform. These include the development of robust and stable cell lines that maintain high productivity over prolonged time periods (e.g., 2-3 months), design of media formulations to support high cell density, and optimization of bioreactor conditions to provide the high cell density at high viability and positive growth rates. Various enabling technologies, such as automatic cell density control, efficient oxygenation and ventilation, and foam control, must be developed and optimized. Significant development is still needed for continuous downstream purification and drug product formulation technologies. For example, although laboratory- and pilot-scale continuous chromatography systems exist, scale-up remains a challenge. Novel downstream unit operations are also needed, such as continuous viral inactivation and removal. An important design requirement for all continuous downstream unit operations is that they must operate over prolonged periods under stringent sterility control conditions, which will require novel and unconventional solutions.

The seamless integration of real-time analytics throughout the synthesis and downstream processes is needed for reliable, long-term continuous operation adhering to tight quality specifications. Physicochemical parameters such as UV absorbance, pH, conductivity, flow rate, pressure, and temperature are reliably monitored in continuous systems using technologies that are well established in batch processing. On the other hand, biotechnology processes require additional process analytical technologies to provide real-time product quality attributes (e.g., activity, aggregation, glycosylation, impurity level). In fact, bioprocessing equipment, instrumentation, and sensors for monitoring, controlling, and sampling are often customized to meet product-specific needs, restricting their widespread adoption in other continuously manufactured product lines.

Whereas localized autonomous systems are sufficient for individual unit operations in batch processing, continuous manufacturing integrates unit operations over the entire process chain, requiring global coordination by a supervisory control system. Such a system must be able to perform oversight functions such as process monitoring, process optimization, and exception handling. The software and hardware systems should provide a high degree of automation and require minimal operator involvement. Standards for aligning local and global control with an improved user interface and the capability for integration with new unit operations will also facilitate continuous manufacturing.

Although the continuous input of an active pharmaceutical ingredient in the manufacturing of a drug product formulation has seen some success (as has the production of biotechnology products by means of continuous perfusion bioreactors), true end-to-end continuous manufacturing from reagents to drug product at a commercial scale has yet to be realized. Economic drivers also need to be considered, as the current inventory of available batch manufacturing facilities may discourage investment in continuous manufacturing.

Federal Investments

While promising at the laboratory scale, the technologies and equipment enabling continuous manufacturing at the commercial scale are not widely available or accessible. Consequently, continuous manufacturing technology is currently in need of further interest and resources from industry, government, and regulatory authorities to translate proofs-of-concept to widespread commercial adoption. Future Federal investments may focus on broadly applicable and enabling technologies, demonstrating their return-on-investment required for mainstream adoption. Process equipment and instrumentation necessary for the continuous manufacturing of large molecules will likely come from smaller, non-traditional biotechnology suppliers seeking to expand their market share, and therefore could be incentivized by government support.

Table 5. Selected examples of Federal investment in continuous manufacturing of pharmaceuticals.

Lead(s)	Title	Scope
NSF	Engineering Research Center for Structure Organic Particulate Systems	This effort aims to build the fundamental science base for the rational design, development and manufacturing of structured organic composite particulate products for the pharmaceutical, food, and agrochemical industries. (TRL 1-4/MRL 1-4) [A]
DOC NIST	Biomanufacturing Program	The NIST Biomanufacturing Program supports the U.S. biopharmaceutical industry through the delivery of standards, measurement science, and tools to enable the more efficient characterization and manufacture of protein-based drugs. (TRL 2-5/MRL 2-5) [B]
DoD DARPA	Pharmacy On Demand and Biologically-derived Medicines On Demand	These efforts aim to enable timely response to patient needs and emergent threats by developing miniaturized and distributed manufacturing platforms to produce multiple small-molecule active pharmaceutical ingredients and protein therapeutics on demand. (TRL 3/MRL 3). [C]
BARDA	Development Through Phase 3: Novel antiviral for the Treatment of Influenza	Exploratory studies to incorporate continuous manufacturing technology into Janssen products under contract to Biomedical Advanced Research and Development Authority (BARDA). (TRL 7/ MRL 8) [D]*
BARDA	Advanced Development of Rempex Carbavance	Exploratory studies to incorporate continuous manufacturing technology into Carbavance manufacturing process under contract to BARDA. (TRL 7/ MRL 8) [E]*
FDA CDER	Advanced Research and Development of Regulatory Science	The Center for Drug Evaluation and Research (CDER) approved the first pharmaceuticals produced by continuous manufacturing and by additive manufacturing, and recently issued a broad-area announcement with BARDA to support emerging and enabling technologies for continuous manufacturing. (TRL 5/MRL 6 for emerging technologies, TRL 7/MRL 8 for enabling technologies) [F]
FDA CDER	Development of Modeling Tools for Quality Risk Management	This effort aims to develop process simulation and modeling platforms for integrated pharmaceutical manufacturing processes. Specifically, the project will focus on manufacturing processes for solid based drug products, which comprise the majority of pharmaceutical products. The tools developed during this project will be used to facilitate the risk assessment of manufacturing processes and control strategies. (TRL 6/MRL 7) [G]

[A] http://www.nsf.gov/awardsearch/showAward?AWD_ID=0540855, http://erc-assoc.org/content/erc-structured-organic-particulate-systems , and http://ercforsops.org/

[B] http://www.nist.gov/mml/bmd/biomanufacturing.cfm

[C] http://www.darpa.mil/program/battlefield-medicine

[D] http://www.hhs.gov/about/news/2015/09/28/hhs-sponsors-development-of-drug-for-hospitalized-influenza-patients.html#

[E] http://www.hhs.gov/about/news/2014/02/05/hhs-funds-drug-for-bioterrorism-antimicrobial-resistant-infections.html

[F] http://www.fbo.gov/spg/HHS/FDA/DCASC/FDABAA-15-00121/listing.html

[G] https://grants.nih.gov/grants/guide/rfa-files/RFA-FD-14-083.html

* The candidate products are at an HHS technology readiness level of 6/7 according to the following website (https://www.medicalcountermeasures.gov/federal-initiatives/guidance/integrated-tris.aspx) and estimated to correspond to 7/8 according to the TRL/MRL table found in Appendix B.

The Federal Government coordinates and fosters a logical progression of technology maturity towards commercial adoption in the field of continuous manufacturing of pharmaceuticals. Federal investments in the basic and emerging science and technology for continuous manufacturing are provided by the National Science Foundation, National Institutes of Health, National Institute of Standards and Technology, and Department of Defense. These investments are well positioned for continued evolution from proof-of-concept to TRL 6 or higher, which can then be transitioned within the private sector in collaboration with the Biomedical Advanced Research and Development Authority (BARDA) for the continued advanced development of BARDA priority medical countermeasures towards regulatory approval. Additionally, these Federal investments serve as workforce development opportunities for scientists and engineers who shepherd technologies through the transition into commercial continuous manufacturing. The Food and Drug Administration (FDA) plays an integral regulatory role from early product development through drug approval. In this role, the FDA provides not only the regulatory guidance and oversight, but also collaborates with drug developers to evolve the regulatory paradigm from batch-centric to continuous manufacturing-centric using scientific and data-driven methodologies to facilitate future approvals, thereby encouraging commercial adoption of continuous manufacturing.

Manufacturing Technology Areas of Existing Priority

In addition to the priority emerging technology areas highlighted elsewhere in this document, there are a number of topics which are the subject of Federal agency interest that have been announced publicly, either for potential future investment or as the focus of an existing Manufacturing Innovation Institute.

Areas of Interest for Potential Future Investment

The following announcements have been made by agencies as areas of interest for potential future Federal support. Please see the weblinks for additional information about these technology areas and their current status.

- NIST released a Federal Funding Opportunity notice to establish and operate one or more National Network for Manufacturing Innovation (NNMI) Institutes.[32] As an open-topic competition, NIST will accept applications for a proposed Institute in any area of advanced manufacturing, so long as it does not duplicate the technical scope and programs of current Institutes within the NNMI or technical focus area(s) of on-going Institute funding opportunities. Like the other institutes that are part of the NNMI, the NIST institutes will be private-public partnerships that foster collaboration among industry, academia, nonprofits and government agencies. Through the institutes, these stakeholders will work to accelerate U.S. innovation and increase competitiveness by investing in industrially relevant, cross-cutting advanced manufacturing technologies and processes.

- DoD issued a Request for Information[33] to consider input from industry, academia, prospective non-profit organizations, and other stakeholders as part of an effort to select and scope the technology focus areas for as many as two more Manufacturing Innovation Institutes. Each institute is launched with Federal funding typically in the range of $70 million to $120 million, leveraging a minimum of 1:1 non-Federal co-investment, over a five to seven years. Technical areas of interest include:

    o Advanced Machine Tools and Control Systems

    o Assistive and Soft Robotics

    o Bioengineering for Regenerative Medicine

    o Bioprinting across Technology Sectors

    o Certification, Assessment and Qualification

    o Open topic (Request for Information responders may suggest)

    o Securing the Manufacturing Digital Thread – Cybersecurity for Manufacturing

- The DOE Advanced Manufacturing Office has recently completed a series of workshops to identify potential topics for future investments, including up to 2 new Manufacturing Innovation Institutes (with anticipated solicitation release in early 2016). The technical areas covered in the workshops are:

---

[32] http://www.nist.gov/amo/nnmi/2016competition.cfm

[33] The solicitation number is RFI-AFRL-RQKM-2016-0009 and can be viewed on the FedBizOps website: http://www.fbo.gov/spg/USAF/AFMC/AFRLWRS/RFI-AFRL-RQKM-2016-0009/listing.html

- ○ Chemical and Thermal Process Intensification[34]
- ○ Sustainability in Manufacturing[35]
- ○ High Value Roll-to-Roll Manufacturing[36]
- ○ Materials for Harsh Service Conditions[37]

Foci of the Manufacturing Innovation Institutes

The following technology areas have already been the subject of significant Administration effort, such as through the existing Manufacturing Innovation Institutes that make up the NNMI (as of February 2016). The Federal Government intends to continue supporting these areas along the technology development pipeline, in order to leverage the distinctive innovation capabilities to-date for the advancement of agencies' missions and overall U.S. industry competitiveness.

Additive Manufacturing

Additive Manufacturing is defined as a process of joining materials to make objects from three-dimensional model data, usually layer upon layer, as opposed to subtractive manufacturing methodologies. Additive manufacturing can encompass metals, polymers, and electronics. Additive manufacturing can apply to a range of structural and functional materials and to a range of components for defense and energy applications. An advantage of additive manufacturing is that parts can be fabricated as soon as the three-dimensional digital description of the part is created, thus establishing a new market for on-demand, mass customization manufacturing. Most importantly, these processes minimize material waste and tooling requirements, as well as drastically compress the supply chain. In addition, novel components and structures can be produced from additive manufacturing processes that cannot be cost effectively produced from conventional manufacturing processes such as casting, molding, and forging.

Established in partnership with DoD and headquartered in Youngstown, OH, America Makes is the National Additive Manufacturing Innovation Institute serving as the national accelerator for additive manufacturing and 3D printing. This Institute is developing the standards, tools, education, and research required to accelerate the U.S. manufacturing sector into a dominant, global economic force. Specifically, five technical areas are addressed in the America Makes Additive Manufacturing Technology Roadmap: design, material, process, value chain, and additive manufacturing genome.

Additional information can be found at: https://americamakes.us/about/overview

Advanced Composites

Lightweight, high-strength, and high-stiffness composite materials have been identified as a key crosscutting technology in U.S. clean energy manufacturing, with the potential to reinvent an energy efficient transportation sector, enable efficient power generation, and increase renewable power production. Priority challenges include: high costs; low production speeds (long cycle times); high manufacturing energy intensity of composite materials; recyclability; and a need to improve design,

---

[34] http://energy.gov/eere/amo/downloads/process-intensification-workshop-september-29-30-2015

[35] http://energy.gov/eere/downloads/workshop-sustainability-manufacturing-january-6-7

[36] http://energy.gov/eere/amo/downloads/workshop-high-value-roll-roll-hv-r2r-manufacturing-innovation-december-2-3-2015

[37] http://energy.gov/eere/amo/downloads/workshop-materials-harsh-service-conditions-november-19-20-2015

modeling, and inspection tools for composites to meet commercial and regulatory demands.

To accelerate the technological advances and research in manufacturing needed to reach cost and performance targets—from constituent materials production to final composite structure fabrication—the President announced the Institute for Advanced Composite Materials Innovation (IACMI). Supported by DOE's Advanced Manufacturing Office and focused on research, development, and demonstration programs, IACMI brings together industry, research institutions, and state partners committed to accelerating development and adoption of manufacturing technologies for low-cost, energy-efficient manufacturing of advanced polymer composites.

Additional information can be found at: http://iacmi.org/about-us/

Digital Manufacturing and Design

Digital manufacturing is the use of an integrated, computer-based system comprised of simulation, three-dimensional visualization, analytics and various collaboration tools to create product and manufacturing process definitions simultaneously. Design innovation is the ability to apply these technologies, tools, and products to re-imagine the entire manufacturing process from end to end.

The Digital Manufacturing and Design Innovation Institute (DMDII), headquartered on Chicago's Goose Island, is in a public-private partnership with the DoD, acting as a first-of-its-kind manufacturing applied research and development lab.

DMDII focuses on enterprise-wide utilization of the digital thread, enabling highly integrated design and manufacturing of complex products in order to reduce time and cost, as well as accelerate the pace of new products coming to market. Processes developed within DMDII that create an open and collaborative environment will help retain supply chain knowledge and improve capabilities to produce low volume, complex systems.

Additional information can be found at: http://dmdii.uilabs.org/the-institute/vision

Flexible Hybrid Electronics

Flexible hybrid electronics preserve the full operation of traditional electronic circuits, but in novel flexible architectures and form factors that are conformal, allowing for bending, stretching, or folding. These highly functional devices can be attached to curved, irregular, and often stretched objects, and have the potential to expand traditional electronic packaging to new form factors, enabling new classes of commercial and DoD technologies. Examples include medical devices and sensors, sensors to monitor structural or vehicle performance, sensors interoperating through the Internet or as sensor clusters to monitor physical positions, wearable performance or information devices, robotics, human-robotic interface devices, and lightweight human-portable electronic systems.

In support of this technology area, the DoD established NextFlex, an end-to-end, sustainable Manufacturing Innovation Institute focused on flexible hybrid electronics manufacturing technology, headquartered in San Jose, CA. The Institute's core focus is on manufacturing assembly and integration, system integration demonstrations, innovative printing processes, thin device processes, materials manufacturability and scale-up, modeling and design tools, and education and training.

Additional information can be found at: http://www.nextflex.us/nnmi/

Integrated Photonics

Integrated photonics encompass the integration of multiple lithographically-defined photonic and electronic components and devices (for example, lasers, detectors, waveguides and passive structures, modulators, electronic controls, and optical interconnects) on a single substrate with nanoscale features. The key benefits of integrating these components include simplified system design, enhanced systems performance, reduced component space and power consumption requirements, and improved component performance and reliability while enabling new capabilities and functionality with lower costs. Today's integrated photonics manufacturing community is made up of a collection of interrelated, largely independent businesses, organizations, and activities that together form an ecosystem, but one that currently lacks the organization and aggregated market strength needed to efficiently innovate manufacturing technologies for cost-effective design, fabrication, testing, assembly, and packaging of integrated photonic devices.

To address this, the DoD established the American Institute for Manufacturing Integrated Photonics, headquartered in Rochester, NY. Reflecting industry priorities, the Integrated Photonics Manufacturing Innovation Institute's principal concerns will be initially five critical topics: computer-aided design tool development; chip fabrication process design kit development; automated testing, assembly, and packaging; prototype demonstrations; and workforce development through focused education and training.

Additional information can be found at: http://www.aimphotonics.com/overview/

Lightweight Metals

Lightweight and modern metals offer significant system performance enhancements and greater energy efficiency. The availability of advanced lightweight metals is a pervasive factor in improving the performance of many systems in defense, energy, transportation and general engineered products, each representing large sectors of the U.S. economy. Moreover, lightweight metals have additional applications in areas such as wind turbines, medical technology, pressure vessels, and alternative energy sources.

The Lightweight Innovations for Tomorrow (LIFT) Institute, headquarter in Detroit, MI, was established through a public-private partnership with the DoD to advance our national fabricated metal product and transportation equipment subsectors. The Institute brings together the Federal and State Governments, leading manufacturers, professional societies and organizations, universities, and other research partners. The institute's mission is to serve U.S. manufacturing by acting as the bridge between basic research and final product commercialization of new, advanced lightweight materials and innovative manufacturing technologies and practices.

Additional information can be found at: http://lift.technology/

Smart Manufacturing

Smart manufacturing—the convergence of information and communications technologies with manufacturing processes to drive unprecedented real-time control of energy, productivity, and costs across factories and companies—was identified by the President's Council of Advisors on Science and Technology[38] as one of the highest-priority manufacturing technology areas in need of Federal investment. Harnessing advanced sensors, controls, information technology processes and platforms, and advanced energy management systems, smart manufacturing has the potential to drive energy efficiency and U.S. manufacturing competitiveness in a range of sectors.

The Advanced Manufacturing Office at DOE released a Request for Information in October 2015 to stand

---

[38] President's Council of Advisors on Science and Technology. *Report to the President: Accelerating U.S. Advanced Manufacturing*. October 2014.
https://www.whitehouse.gov/sites/default/files/microsites/ostp/PCAST/amp20_report_final.pdf

up a Clean Energy Manufacturing Innovation Institute around the topic of Smart Manufacturing—the 9th Institute in the National Network for Manufacturing Innovation. Representing roughly $70 million in Federal investment, this public-private partnership will address mid-TRL R&D needs for the software and physical hardware necessary to support Smart Manufacturing, as well as soft barriers like education, supply chain dissemination, integration with existing infrastructure, and workforce.

Additional information can be found at: http://www.energy.gov/articles/energy-department-announces-70-million-innovation-institute-smart-manufacturing

Revolutionary Fibers and Textiles

Scientific advances have enabled fibers and textiles with extraordinary properties including strength, flame resistance, and electrical conductivity. These types of fibers and textiles are composed of specialty fabrics, industrial fabrics, electronic textiles, and advanced textiles. These technical textiles are built upon a foundation of synthetic, natural fiber blends and/or multi-material fibers that have a wide range of applications, in both the defense and commercial sectors, which go beyond traditional wearable fabrics.

The DoD announced the competition for a Revolutionary Fibers and Textiles Manufacturing Innovation Institute on March 18, 2015. The Federal Government expects the Revolutionary Fibers and Textiles Manufacturing Innovation Institute to support an end-to-end ecosystem in the United States for revolutionary fibers and textiles manufacturing and include domestic manufacturing development facilities to scale up manufacturing processes. The Revolutionary Fibers and Textiles Manufacturing Innovation Institute positions the domestic industrial base to recapture and secure U.S. leadership in the technical and smart-textile marketplace.

Additional information can be found at: http://www.manufacturing.gov/rft-mii.html

Wide Bandgap Electronics

Wide bandgap semiconductor materials allow power electronic components to be smaller, faster, more reliable, and more efficient than their silicon-based counterparts. These capabilities make it possible to reduce weight, volume, and life-cycle costs in a wide range of power applications. If widespread adoption of these technologies is realized in even a limited set of applications, then 40,100 GWh (137 TBtu) of electrical power savings in the United States could be achieved annually. With higher volume production, the high cost of wide bandgap substrate and epitaxial deposition, which is currently tied to small production volumes and high manufacturing costs, is expected to decrease.

As part of the NNMI, the DOE Advanced Manufacturing Office is currently supporting PowerAmerica. Led by North Carolina State University, PowerAmerica focuses on making wide bandgap semiconductor technologies cost-competitive with the silicon-based power electronics that are currently used. The Institute is establishing a collaborative community that will create, showcase, and deploy new power electronic capabilities, products, and processes that can impact commercial production, build workforce skills, enhance manufacturing capabilities, and foster long-term economic growth in the region and across the nation. Additional information can be found at: http://www.poweramericainstitute.com/

Manufacturing Education and Workforce Training Priorities

Description

U.S. manufacturing industries employ over 12 million people[39] and have diverse workforce needs, drawing skillsets from engineering, computer science, biology, mathematics, economics, and psychology, among other disciplines. Manufacturing requires highly-skilled craftsmen, technicians, designers, planners, researchers, engineers, and managers, whose specialized educations span the spectrum from theory to practice.

Manufacturing is typically carried out on expensive, specialized equipment that is often impractical to accurately duplicate or simulate in an educational setting. Therefore, it is important for manufacturing industries to play an active role in the educational process to ensure proper training for the next generation of their workforce. Research collaborations connecting engineers from industry with university students, involvement of companies in equipping laboratories and designing curricula for community college programs, and hosting "how it's made" field trips to factories for elementary and secondary school students all help to bridge from theory to practice and orient students towards the actual needs of the companies for which they will be working.

While manufacturing jobs are high-paying, high-technology jobs, individual career choices are often based on an obsolete view of manufacturing. This dynamic puts a premium on providing educational experiences at every level of education, from primary school to postgraduate, that accurately reflect the challenges and opportunities of advanced manufacturing.

Federal Investments

Federal agencies have innovated education and workforce training programs to address these key challenges, and resulting investments range from specialized technician training at community colleges to student education in graduate research programs at world-leading research universities.

The NSF's Advanced Technological Education program supports collaborations between academic institutions and industry to improve the education of technicians for high-technology fields at two-year community and technical colleges, with components that reach down to the secondary school and up to the undergraduate university levels. The program, launched in 1993 in response to the Scientific and Advanced Technology Act of 1992, currently supports seven advanced manufacturing Advanced Technology Education Centers.[40] Advanced Technology Education Centers have developed curricular materials and skills standards in collaboration with industry, provided professional development for secondary school teachers and both 2-year and 4-year faculty, and developed clearly articulated career pathways into the supported industries. For example, The Florida Advanced Technological Education Center links employers with a network of ten community and state colleges across Florida. The Center's state-wide Engineering Technology Associate Degree Program focuses on a set of core courses covering introductory computer-aided drafting, electronics, instrumentation and testing, processes and materials, quality, and safety, and integrates industry validated stackable credentials into a 2-year technician training curriculum. Over 140 companies have participated in these education and outreach activities since the founding of the center in 2004.

The Manufacturing Innovation Institutes of the NNMI connect talented and experienced manufacturing professionals with upcoming generations of students to provide real-world training for their future

---

[39] www.bls.gov/iag/tgs/iag31-33.htm#workforce

[40] http://www.nsf.gov/funding/pgm_summ.jsp?pims_id=5464 and www.atecenters.org/advanced-manufacturing-technologies

manufacturing workforce. For example, the Lightweight Innovations for Tomorrow Manufacturing Innovation Institute has launched an advanced manufacturing apprenticeship learning pathways model with Jefferson Community and Technical College in Louisville, Kentucky. This new initiative will both develop a pipeline of trained entry-level workers and implement an accelerated, modularized and skills-based program for those in the current workforce who are in need of continuing education and training. Many local companies, from Fortune 100 companies to small and medium-sized manufacturers, have expressed interest in training and hiring apprentices through the program. The Institute has also established the Right Skills NOW program for computer numerical control machinist training with Vincennes University. This accelerated program of study transitions veterans to jobs as skilled computer numerical control machinists and incorporates national credentials from the National Institute for Metalworking Skills. Veterans participating in the Right Skills NOW program will train at the newly built Gene Haas Training Center, a state-of-the-art, 23,000 square foot facility. Partner companies mentor trainees during training and graduates are immediately placed in positions in industry after program completion.

The Departments of Labor and Education have deployed nearly $1 billion through the Trade Adjustment Assistance Community College and Career Training (TAACCCT) grant program[41] to strengthen manufacturing curricula at community colleges across the country and train America's workforce. Funded programs range from short-term certificate programs to associates degrees and many of the TAACCCT programs offer stackable credentials that provide skills for rapid entry in to the workforce and upgrading skills on the job. Manufacturing subsectors addressed include medical devices, additive manufacturing and3D printing, advanced manufacturing, advanced materials, robotics and mechatronics, and more traditional occupational areas, such as computer numerical control machining and welding. Henry Ford Community College, in Dearborn, Michigan, is the lead of the Multi-State Advanced Manufacturing Consortium project and provides an example of the numerous TAACCCT manufacturing collaborations currently under way. This project will establish a model that is applicable to many industries. The strategy incorporates a competency-based, industry-driven, credentialed manufacturing curriculum; new instructional design and delivery systems to accelerate and contextualize learning; improved student support, success, and placement strategies; and administrative structures to support instructional redesign. Five of the top seven global automobile manufacturers and other major manufacturers are collaborating in the program.

The Jobs and Innovation Accelerator Challenge in Advanced Manufacturing—a joint program of the Departments of Labor, Commerce, and Energy and the Small Business Administration—supports local efforts to spur job creation by connecting innovative small suppliers with large companies, linking research with start-ups that can commercialize new ideas, and providing trained workers with the skills they need to help their companies capitalize on business opportunities. Similarly, the Departments of Commerce and Labor and the Delta Regional Authority support the Make It In America grant program. Make It In America grants encourage businesses to build and/or expand their operations in the United States through multiple mechanisms, including targeted training and employment activities that support the skill-building and career advancement needs of U.S. workers, thereby strengthening America's highly-skilled and diverse workforce. For example, Make It In America grantee South Carolina Appalachian Council of Governments supports workforce development, economic development, and job creation through electronic access to industry-recognized, high-technology educational programs using a variety of electronic tools, increasing efficiency in industry supply chain management and Innovation Engineering and increasing human capital capacity through training activities with embedded industry credentials.

---

[41] https://www.doleta.gov/taaccct/

41

South Carolina Appalachian Council of Governments is connecting current and former students with interested employers through networks formed by instructors, corporate representatives, and other community contacts.

More recently, the Department of Labor invested $175 million through the American Apprenticeship Initiative grant competition to train and hire more than 34,000 new apprentices in high-growth and high-technology industries in the next five years. 27 of the 46 American Apprenticeship Initiative grants include advanced manufacturing as a targeted sector.

The NSF's Industry/University Cooperative Research Centers (I/UCRC) Program[42] and the Engineering Research Center (ERC) Program[43] bring an industry-focused orientation to academic research that enables students to be productive employees from their first day on the job. I/UCRC Program graduate and undergraduate students perform precompetitive research on industry-validated projects in multi-member, sustained collaborations between industry, academia, and government. NSF provides a financial and procedural framework for membership and operations, in addition to best practices learned over decades of fostering public-private partnerships. As an example, one of the ten advanced manufacturing Centers currently supported is the I/UCRC Program on Intelligent Maintenance Systems, a partnership with over 40 current member companies and the Universities of Cincinnati, Michigan and Texas and the Missouri University of Science and Technology. This partnership utilizes advanced sensing, the Internet-of-Things, and data analytics to enable products and systems to achieve near-zero breakdown performance. Funding is largely provided by industry and the project agenda is jointly determined by Center leadership and industry members, with over 100 projects conducted over the life of the Center. According to the 2012 *Measuring the Economic Impacts of the NSF Industry/University Cooperative Research Centers Program: A Feasibility Study*,[44] the Center delivered its members over $800 million in combined benefits over its first ten years.

Whether in or out of school, significant learning takes place in informal settings where access to learning tools and mentors accelerates independent study. Recognition of the importance of bringing this modality to manufacturing is growing, including White House Maker Faires[45] that highlighted the importance of the maker movement and demonstrated today's do-it-yourself as tomorrow's "Made in America." In recent years, the affordability and capabilities of desktop manufacturing equipment and the availability of makerspaces have all increased, making it easier for people to design and fabricate products and bring them to market. Individuals are self-identifying as makers and, in doing so, are forming communities of interest that facilitate the sharing of technical information and launching of crowdfunding campaigns to scale production of their products and become entrepreneurs. Investments in research to develop the next generation of low-cost tools and technologies for design and fabrication will be required to continue to advance the variety of things that individuals can create and produce in the United States. Finding innovative approaches to help makers scale more easily from small batch manufacturing to larger volumes will be critical to the growth and sustainability of these business in the United States.

---

[42] http://www.nsf.gov/eng/iip/iucrc/program.jsp

[43] http://www.nsf.gov/funding/pgm_summ.jsp?pims_id=5502

[44] https://www.ncsu.edu/iucrc/PDFs/IUCRC_EconImpactFeasibilityReport_Final.pdf

[45] www.whitehouse.gov/nation-of-makers

Among such efforts, NSF's Cybermanufacturing research and educational awards[46] are helping to define the requirements for an application-based manufacturing infrastructure that can be accessed by makers through the web. University classrooms and laboratories provide a nearly ideal environment for researching, developing and exercising such innovations. In addition, DARPA's Manufacturing Experimentation and Outreach Two program enhances defense readiness by developing and demonstrating new manufacturing training, tools and materials for those who will be called on to use, maintain, and adapt high-technology systems in low-technology environments. DARPA envisions that project-based curricula employing this program's design and prototyping tools will teach a deeper understanding of high-technology systems and better enable future competence in the maintenance and adaptation of such systems through the manufacture of as-designed components or the design and manufacture of new components. Through this program, for example, DARPA and the U.S. Navy installed a pilot fabrication laboratory, or Fab Lab, at the Mid-Atlantic Regional Maintenance Center in Norfolk, Virginia, working with the Fab Foundation.[47] These programs are examples of how Federal agencies are leveraging existing assets and resources and investing in the development of new tools and technologies to democratize design, prototyping, and manufacturing skills and capabilities for individuals throughout the United States.

---

[46] http://nsf.gov/pubs/2015/nsf15061/nsf15061.jsp

[47] https://www.whitehouse.gov/blog/2015/12/10/makers-military

Conclusions

U.S. advanced manufacturing supports long-term economic prosperity, security, and growth, as well as the missions of the Federal agencies of the Subcommittee for Advanced Manufacturing. To unleash the advanced manufacturing industries of the future, and ensure U.S leadership in these industries, concerted efforts of government, industry, and academia must focus on the key technologies that underpin U.S. competitiveness. In particular, collaborative pre-competitive research and development should leverage common assets to the benefit of all stakeholders, overcoming the significant development cycle costs that are impractical for any single entity to bear alone.

This report provides a collection of emerging technology areas in advanced manufacturing that are priorities for Federal members of the Subcommittee for Advanced Manufacturing and that are strong candidates for focused Federal investment and public-private collaboration. These areas include:

- Advanced materials manufacturing

- Engineering biology to advance biomanufacturing

- Biomanufacturing for regenerative medicine

- Advanced bioproducts manufacturing

- Continuous manufacturing of pharmaceuticals

This report further captures a number of existing technology areas, which are already the topics of substantial existing investments (such as the Manufacturing Innovation Institutes) or that may be the subject of future programming.

In recognition of the critical role of education and workforce training in building and sustaining Tomorrow's advanced manufacturing industries, an array of Federal programs that encourage strong industry involvement are described.

Appendix A. Federal Participants in the NSTC SAM

The National Economic Council

The National Economic Council was established in 1993 to advise the President on U.S. and global economic policy. The National Economic Council has four principal functions: to coordinate policy-making for domestic and international economic issues, to coordinate economic policy advice for the President, to ensure that policy decisions and programs are consistent with the President's economic goals, and to monitor implementation of the President's economic policy agenda. More information is available at https://www.whitehouse.gov/administration/eop/nec.

Office of Management and Budget

More information is available at https://www.whitehouse.gov/omb/.

The Office of Science and Technology Policy

The Office of Science and Technology Policy (OSTP) was established by the National Science and Technology Policy, Organization, and Priorities Act of 1976. OSTP's responsibilities include advising the President in policy formulation and budget development on questions in which science and technology are important elements; articulating the President's science and technology policy and programs; and fostering strong collaborations among Federal, state, and local governments, and the scientific communities in industry and academia. The Director of OSTP also serves as Assistant to the President for Science and Technology and manages the National Science and Technology Council (NSTC). More information is available at https://www.ostp.gov.

Department of Agriculture

Worldwide, the bioenergy and bioproducts industries are emerging as new and rapidly growing sectors; given the high productivity of the U.S. agricultural sector, biobased product manufacturing is a significant opportunity for the United States to support growth of a bioeconomy. Expansion of the bioeconomy has the potential to sustainably harvest and utilize one billion tons of new biomass in the United States without affecting existing farm and forestry product markets, growing the current market five-fold over the next fifteen years and adding $500 billion to the annual bioeconomy.

The agricultural sector is essential for ensuring sustainable, reliable, and accessible production of bioenergy and biobased products that: (1) replace the use of petroleum and other strategic materials that would otherwise need to be imported; (2) create higher-value revenue streams for producers in rural and agricultural communities; (3) improve the nutrition and well-being of animals and humans; and (4) provide ecosystem services such as ensuring clean air and water, biodiversity, and nutrient cycling to the environment and society.

The U.S. Department of Agriculture recognizes the role that manufacturing plays in maximizing the benefits of a sustainable, rural economy. Areas of interest include biomanufacturing and bioproducts development to: (1) establish processes and chemical platforms leading to high-value intermediate and end-use products, (2) support commercialization of products developed from basic and applied research, (3) build domestic capability for ongoing biomanufacturing and bioproducts development, and 4) educate and train needed workforce. The growth of the bioeconomy also depends upon understanding and addressing the entire supply chain of the bioeconomy, rural America's role in the bioeconomy, and the role of research and development.

In addition, nanocellulose materials have enormous promise to bring about fundamental changes in and significant benefit from our Nation's use of renewable resources. These cellulose nanomaterials when derived from trees: (1) are renewable and sustainable; (2) are produced in trees via photosynthesis from solar energy, atmospheric carbon dioxide, and water; (3) store carbon; and (4) depending upon how long cellulose-based products remain in service, are carbon negative or carbon neutral. Cellulosic nanocrystals, for example, are predicted to have strength properties comparable to Kevlar, have piezoelectric properties comparable to quartz, and can be manipulated to produce photonic structures. Current global research directions in cellulose nanomaterials indicate that this material could be used for a variety of new and improved product applications, including lighter and stronger paper and paperboard products, lighter and stronger building materials, wood products with improved durability, barrier coatings, body armor, automobile and airplane composite panels, electronics, biomedical applications, and replacement of petrochemicals in plastics and composites.

Department of Commerce

Innovation results from initial advances that lead to additional technology and process improvements, with resulting benefits accruing to industry, the economy, and society as a whole. Innovation in advanced manufacturing begins with the generation of new ideas that are refined and matured through applied research, development, and invention. Manufacturers then scale those ideas for mass production in order to generate process improvements and make new products. The experience and knowledge gained through manufacturing then leads to new ideas that start the cycle again. The Department of Commerce (DOC) has central responsibility for supporting and expanding each part of this cycle and has the relationships with businesses necessary to identify the workforce skills needed to support new and growing industries.

The Department increases regional and national capacity for innovative manufacturing through partnerships with state and local governments, academic institutions, and the private sector. Through the Department's convening power, regional economic development programs, and statistical and economic analysis, it empowers industry-driven solutions to the shortage of high demand skills. Finally, the Department supports research and development leading to transformative changes in technology and promotes intellectual property policy that supports and protects innovation. By supporting public-private partnerships, such as the National Network for Manufacturing Innovation (NNMI), the Department helps to accelerate technology development and commercialization, and strengthen the Nation's position in the global competition for new products, new markets, and new jobs.

National Institute of Standards and Technology

The National Institute of Standards and Technology (NIST) mission is: To promote U.S. innovation and industrial competitiveness by advancing measurement science, standards, and technology in ways that enhance economic security and improve our quality of life. For more than 110 years, NIST has maintained the national standards of measurement, a role that the U.S. Constitution assigns to the Federal Government to ensure fairness in the marketplace. Today, the NIST Laboratories address increasingly complex measurement challenges, ranging from the very small (nanoscale devices) to the very large (vehicles and buildings), and from the physical (renewable energy sources) to the virtual (cybersecurity and cloud computing). As new technologies develop and evolve, NIST's measurement research and services remain central to innovation, productivity, trade, and public safety. NIST promotes the use of measurements based on the international system of units (SI). The measurement science research at NIST is useful to all science and engineering disciplines, including a robust research portfolio targeting the specific technical challenges associated with advanced manufacturing. In addition, the NIST Manufacturing Extension Partnership (MEP) is a critical resource to engage small- and medium-sized manufacturers to develop new products, expand into global markets, and adopt new technologies, such as those in development in the NNMI Manufacturing Innovation Institutes. More information is available at http://nist.gov/.

The Advanced Manufacturing National Program Office

Hosted at NIST, the Advanced Manufacturing National Program Office (AMNPO) is an interagency team with participation from Federal agencies involved in advanced manufacturing. Principal participant agencies currently include the Departments of Commerce, Defense, Education, and Energy, the National Aeronautics and Space Administration, and the National Science Foundation. Established in 2012, the Advanced Manufacturing National Program Office reports to the Executive Office of the President and operates under the NSTC on cross-agency initiatives. The office reports to the Secretary of Commerce in its role as the "the National Office of the Network for Manufacturing Innovation Program," also referred to as the "National Program Office," as described by the Revitalize American Manufacturing and Innovation Act of 2014.[48] More information is available at http://www.manufacturing.gov.

Department of Defense

The Department of Defense (DoD) requires a mechanism for shaping and developing the domestic design and manufacturing industrial base in support of national security needs. There is a need to advance the maturity of manufacturing processes in order to bridge the gap from research and development to full-scale production and aid in the economical and timely acquisition of weapon systems and components. New emerging technologies hold strategic promise for the DoD, but their fragmented and frail ecosystems are at risk of collapse due to infrastructure and workforce complexities. An ecosystem established for DoD requirements only is insufficient to establish a robust and sustainable ecosystem, therefore advanced manufacturing ecosystems must be built on common commercial and defense manufacturing and design challenges for shared risks and shared benefits.

The DoD Manufacturing Innovation Institutes, a key investment strategy for DoD, are designed to overcome many of these challenges by advancing manufacturing innovation for specific, focused technology area manufacturing ecosystems. As of March 2016, there are five DoD Institutes in operation for additive manufacturing, digital manufacturing, lightweight metals, integrated photonics, and flexible hybrid electronics. A sixth for revolutionary fibers and textiles is scheduled for award early this year and two additional Manufacturing Innovation Institutes are planned over the next two years.

Department of Education

The Department of Education supports education at all levels with across-the-board relevance to the knowledge and skill needs of the economy. Particular programs and initiatives focus on science, technology, engineering, and mathematics fields, which are especially important in building the technically skilled workforce needed by the manufacturing sector. Most significantly, the Department administers funds that support career and technical education programs in local education agencies and community colleges across the nation. Further, the Department conducts leadership and technical assistance activities to promote quality career and technical education programs that are well articulated between secondary and postsecondary levels, and lead to successful careers. A particular focus for leadership and assistance programs is on advanced manufacturing, and the Department is supporting Federal efforts to revive this sector through its support for the technical skills agenda.

---

[48] Consolidated and Further Continuing Appropriations Act, 2015, Title VII — Revitalize American Manufacturing and Innovation Act of 2014, codified at 15 U.S.C. § 278s.

The Department has been active in helping develop NNMI from its formation, and collaborates with other Federal agencies, in particular those that focus on the knowledge and skill needs of the economy and efforts related to student success.

Department of Energy

The DOE mission is to ensure America's security and prosperity by addressing its energy, environmental and nuclear challenges through transformative science and technology solutions. This includes catalyzing the timely, material, and efficient transformation of the nation's energy system and securing U.S. leadership in clean energy technologies, as well as maintaining a vibrant U.S. effort in science and engineering as a cornerstone of our economic prosperity. To accomplish these goals, the DOE has established the Clean Energy Manufacturing Initiative as a cross-cutting initiative within the Department to strengthen U.S. clean energy manufacturing competitiveness and to increase U.S. manufacturing competitiveness across the board by boosting energy productivity and leveraging low-cost domestic energy resources and feedstocks. Clean energy manufacturing involves the minimization of the energy and environmental impacts of the production, use, and disposal of manufactured goods, which range from fundamental commodities such as metals and chemicals to sophisticated final-use products such as automobiles and wind turbine blades. The manufacturing sector, a subset of the industrial sector, consumes 25 exajoules (24 quads) of primary energy annually in the United States—about 79% of total industrial energy use. The DOE partners with private and public stakeholders to support the research, development, and deployment of innovative technologies that can improve U.S. competitiveness, save energy, and ensure global leadership in advanced manufacturing and clean energy technologies.

The DOE uses Manufacturing Innovation Institutes, which are executed through the Advanced Manufacturing Office, to develop energy efficiency and renewable energy technologies to support the Clean Energy Manufacturing Initiative. To date, the DOE has awarded two Manufacturing Innovation Institutes. The first, PowerAmerica, is focused on wide bandgap semiconductor technologies for next generation power electronics. The second, the Institute for Advanced Composites Manufacturing Innovation, is focused on composite technologies for vehicles, wind turbine blades, and compressed gas storage tanks. A third Institute, Smart Manufacturing: Advanced Sensors, Controls, Platforms and Modelling for Manufacturing, has been issued as an open competitive solicitation at the time of this report.

The DOE led Institute topics have been identified through extensive public engagement and are aligned with the DOE and EERE mission in clean energy science and technology.[49] Specific topical areas are further informed by the DOE Quadrennial Technology Review, particularly the clean energy opportunities for advanced manufacturing.[50]

Department of Health and Human Services

Biomedical Advanced Research and Development Authority

The Biomedical Advanced Research and Development Authority (BARDA), within the Office of the Assistant Secretary for Preparedness and Response in the U.S. Department of Health and Human Services, was established through the enactment of the 2006 Pandemic and All-Hazards Preparedness Act (PAHPA) and reaffirmed by the 2013 Pandemic and All-Hazards Preparedness Reauthorization Act (PAHPRA), to provide an integrated, systematic approach to the development and purchase of the necessary vaccines, drugs,

---

[49] http://energy.gov/eere/downloads/eere-strategic-plan

[50] http://energy.gov/downloads/chapter-6-innovating-clean-energy-technologies-advanced-manufacturing

therapies, and diagnostic tools for public health medical emergencies.

BARDA'S overarching vision is of a Nation with the capability to respond quickly and effectively to deliberate, natural, and emerging threats so as to minimize their impact and recover promptly. A critical enabling factor for the realization of this vision is the existence of a robust domestic pharmaceutical and biotechnology sector that actively collaborates with the Federal Government to address unmet medical countermeasure and public health requirements.

BARDA's mission is to develop and procure needed medical countermeasures—including vaccines, therapeutics, diagnostics, and non-pharmaceutical countermeasures—against a broad array of public health threats, whether natural or intentional in origin. BARDA was created to address gaps in the Federal Government medical countermeasure development and procurement process and to bridge the "valley of death" that separates candidates identified in early research from potential FDA licensure/approval by providing funding, technical support and services necessary to advance candidate products through the developmental pipeline.

Food and Drug Administration

The U.S. Food and Drug Administration (FDA) is an agency within the Department of Health and Human Services. FDA is responsible for promoting and protecting the public health by assuring the safety and efficacy of human and veterinary drugs, vaccines, blood, and other biological products, medical devices and radiation-emitting products. Issues in pharmaceutical manufacturing have the potential to significantly impact patient care, in that failures in quality may result in product recalls and harm to patients. Additionally, failures in product or facility quality are a major factor leading to disruptions in supply of medicines. Modernizing manufacturing technology may lead to a more robust manufacturing process and greater assurance that the drug products manufactured in any given period of time will provide the expected clinical performance.

The FDA laid out a 21st Century vision to modernize the regulation. To support this vision and mission, the FDA conducts research to support the development of scientific standards and policies in the composition, quality, safety, and effectiveness of human drug products to meet public health needs; identifies, evaluates, and develops relevant technologies to assess the safety and efficacy of human drug products; conducts research studies to address unresolved regulatory science questions and to prepare for issues related emerging technologies; and collaborates with academia, industry, and other government agencies to stimulate development of novel manufacturing and characterization technologies.

To complement its research program, the FDA has created an Emerging Technology Team to serve as the primary point of contact for companies that are interested in implementing emerging manufacturing technology in the manufacture of their drug products. The Emerging Technology Team will answer sponsor/applicant questions about the information FDA expects to see in their submission; identify and help facilitate regulatory review of a new manufacturing technology in accordance with existing legal and regulatory standards, guidance, and agency policy related to quality assessment; serve as the lead or co-lead on the quality assessment team and make the final quality recommendation regarding the potential approval of submissions in the program; and identify and capture resolution to policy issues that may inform FDA approaches and recommendations regarding future submissions that involve the same technology. More information is available in the draft guidance:
http://www.fda.gov/downloads/Drugs/GuidanceComplianceRegulatoryInformation/Guidances/UCM4788 21.pdf

National Institutes of Health

NIH's mission is to seek fundamental knowledge about the nature and behavior of living systems and the application of that knowledge to enhance health, lengthen life, and reduce illness and disability. The goals of the agency are to foster fundamental creative discoveries, innovative research strategies, and their applications as a basis for ultimately protecting and improving health; to develop, maintain, and renew scientific human and physical resources that will ensure the Nation's capability to prevent disease; to expand the knowledge base in medical and associated sciences in order to enhance the Nation's economic well-being and ensure a continued high return on the public investment in research; and to exemplify and promote the highest level of scientific integrity, public accountability, and social responsibility in the conduct of science.

In realizing these goals, the NIH provides leadership and direction to programs designed to improve the health of the Nation by conducting and supporting research in harnessing new technologies to improve health. Rapid expansion of technological capabilities has opened new horizons for biomedical research. Innovative research methods stimulated by technological and engineering advances are facilitating the development of new strategies to diagnose, prevent, and treat a host of diseases. Technology also facilitates the integration of previously disparate fields, such as biology and electronics, enabling NIH to cultivate new lines of medical research and practice.

Department of Homeland Security

The Department of Homeland Security (DHS) seeks to achieve a safe, secure, and resilient homeland. Given the current and projected threat environments, technology and research and development (R&D) are the bridge to the future of homeland security. The most effective and efficient changes will come with the smart application of science and technical expertise to develop force-multiplying solutions. These technology-based solutions will provide homeland security operators and first responders the upper hand in their respective operational spaces. They will also enable the Homeland Security Enterprise to expand capabilities and security coverage. DHS leverages the Integrated Product Team process that aligns mission areas and incorporates a Science and Technology Directorate-led technology assessment for all major acquisitions. Integrated Product Teams are tasked to identify DHS capability gaps and coordinate R&D to close those gaps across departmental mission areas. Additionally, the Science and Technology Directorate's Targeted Innovative Technology Acceleration Network and Responder Technology Alliance programs, in collaboration with the Homeland Security Industrial Base, strive to smartly manufacture and deliver technology solutions for the Homeland Security Enterprise and first responder community.

Each of the five Science and Technology Directorate's Homeland Security Advanced Research Projects Agency divisions and the three First Responders Group divisions, as well as the Apex programs and Technology Engines, have developed Capability Roadmaps aligned to the needs of their operational end users. These high-level roadmaps formalize a vision, identify strategic drivers, and list R&D objectives for the next five years. The roadmaps are constantly evolving documents, serving three primary organizational functions: (1) to build consensus among a diverse set of end users with similar operational requirements; (2) to develop a framework that directly links a strategy to tactics; and (3) to provide a framework to coordinate planning, research, development, and acquisition activities across the various groups involved.

Department of Labor

The Department of Labor's mission is to foster, promote, and develop the welfare of the wage earners, job seekers, and retirees of the United States; improve working conditions; advance opportunities for profitable employment; and assure work-related benefits and rights. The rapidly changing manufacturing workplace has led to the growing demand for workers and concerted efforts to support the sector through national, state, and local economic and workforce development activity. In alignment with its mission, the Department of Labor administers several grant programs that seek to increase the number of skilled workers in the high-demand manufacturing field. These include the Trade Adjustment Assistance Community College and Career Training Grant Program, Jobs Accelerator Innovation Challenge, Make It In America Grant Program, and American Apprenticeship Grant Program. The Department is also supporting in a White House-led initiative, Investing in Manufacturing Communities Partnership, that aims to spur communities to develop integrated, long-term economic development strategies that sharpen their competitive edge in attracting global manufacturers and their supply chains to our local communities—increasing investment and creating jobs.

Department of Transportation

Advances in automotive electronics have enabled many capabilities that have seen more electronics packed into vehicles than ever before. Capabilities like GPS, Wi-Fi, Bluetooth, radar, eye detection systems and infotainment, and more recently, connected vehicles, improve safety and efficiency on the roads and enhance the overall driving experience. The manufacturing of these critical components and ensuring the integrity of the devices as they are made is becoming increasingly important and has come to the attention of the suppliers, as well as their customers and the Department of Transportation.

The stringent requirements of the automotive regulatory environment necessitate strong engineering and validation capability, along with state-of-the-art manufacturing. The current state of manufacturing of automotive electronics does not support component integrity protection. In addition, with much of the production currently done off-shore, ensuring the stringent manufacturing process is proving to be a challenge. The Department of Transportation is exploring a collaborative effort involving automotive electronic component manufacturers to identify ways to better ensure the integrity of the manufacturing process.

Environmental Protection Agency

Under the authorities of the Pollution Prevention Act of 1990, the Environmental Protection Agency (EPA) promotes source reduction to eliminate or reduce pollution at its source. One section in the Pollution Prevention Act of 1990 asks EPA to establish a grant program for states in order to support programs that: (1) make specific technical assistance available to businesses seeking information about source reduction opportunities, including funding for experts to provide onsite technical advice to businesses seeking assistance and to assist in the development of source reduction plans; (2) target assistance to businesses for whom lack of information is an impediment to source reduction; (3) provide training in source reduction techniques. Such training may be provided through local engineering schools or any other appropriate means. EPA's Pollution Prevention grant program reduces millions of pounds of pollution, a million metric tons of carbon dioxide equivalents and nearly a billion gallons of water use every year.

Under the Pollution Prevention Program, EPA also supports U.S. manufacturers seeking to improve their economic performance and competitiveness through improved environmental performance. The Presidential Green Chemistry Challenge Awards promote the environmental and economic benefits of developing and using novel green chemistry. These prestigious annual awards recognize chemical technologies that incorporate the principles of green chemistry into chemical design, manufacture, and use. EPA's Office of Chemical Safety and Pollution Prevention sponsors the Presidential Green Chemistry Challenge Awards, in partnership with the American Chemical Society Green Chemistry Institute and other members of the chemical community including industry, trade associations, academic institutions, and other government agencies. Through 2015, the 104 winning technologies have made billions of pounds of progress, including: 826 million pounds of hazardous chemicals and solvents eliminated each year—enough to fill almost 3,800 railroad tank cars or a train nearly 47 miles long; 21 billion gallons of water saved each year—the amount used by 820,000 people annually; 7.8 billion pounds of carbon dioxide equivalents released to air eliminated each year—equal to taking 810,000 automobiles off the road.

EPA also develops standards, criteria documents, and ecolabeling programs for products as part of its mission to protect human health and the environment. Examples of EPA ecolabeling programs include ENERGY STAR™, WaterSense®, and Safer Choice. They are noteworthy examples of Federal leadership in advancing energy efficiency, water efficiency, and green chemistry, respectively, and reflect EPA's commitment to objective, fact-based decision-making, grounded in scientific reasoning and principles, and using the best available data. The Economy, Energy and Environment Program is a Federal technical assistance framework helping communities, manufacturers, and manufacturing supply chains adapt and thrive in today's green economy. EPA and five other Federal agencies have pooled their resources to support small and medium-sized manufacturers with customized assessments. The Economy, Energy and Environment Program is helping communities across the country reduce pollution and energy use while increasing profits and creating new job opportunities.

National Aeronautics and Space Administration

The National Aeronautics and Space Administration (NASA) depends on manufacturing innovation to enhance its technical and scientific capabilities in aeronautics and space exploration. NASA will support the NNMI Program through funded research and development to help stimulate its mission-related capacity for innovation and economic growth within the government, at universities, and at industrial companies.

NASA's Space Technology Mission Directorate serves as the agency's principal organization supporting the NNMI Program. The Space Technology Mission Directorate rapidly develops, demonstrates, and infuses revolutionary, high-payoff technologies through transparent, collaborative partnerships, expanding the boundaries of the aerospace enterprise. By investing in bold, broadly applicable, disruptive technology that industry cannot tackle today, the Space Technology Mission Directorate seeks to mature the technology required for NASA's future missions in science and exploration while proving the capabilities and lowering the cost for other government agencies and commercial space activities. These collective efforts give NASA the ability to do first-of-a-kind missions and longer-term advancements in research and technology—those beyond what industry will take on and those focused on national advancement in aeronautics and space that also align with NASA's role in the NNMI Program.

NASA will leverage the NNMI Program to support advanced manufacturing technology research and development as a critical means of addressing improved affordability, enhanced performance, and improved safety and reliability for NASA's aerospace research and development efforts. NASA investments span low, medium, and high TRLs through multiple NASA programs including Small Business Innovation Research Program, Small Business Technology Transfer, Space Technology Research Grants, Game Changing Development, Technology Demonstration Missions, and other grant opportunities.

Advanced manufacturing research and development at NASA is focused in several areas: cutting-edge materials, additive manufacturing (3D printing), polymer matrix composites, metals processing/joining, robotics, computational physics-based modeling, non-destructive evaluation, and other highly specialized areas. This research and development is conducted through a combination of in-house activities at NASA centers, competitively funded research with universities and industry, and collaborations with other agencies, universities, and industry. The rapid infusion of advanced manufacturing technologies into mission applications is a major emphasis of NASA's technology investment plan.

NASA is expanding its efforts to engage industry and academia on advanced manufacturing topics central to the nation's space mission through its National Center of Advanced Manufacturing, with a particular focus to develop "technology testbeds" within its research facilities and manufacturing technologies that reduce the weight of materials during space flight.

NASA has participated in the NNMI since its inception and is committed to partnering with other participating agencies to identify key technical challenges in advanced manufacturing research and development, focus resources to address these challenges, and accelerate the development of advanced manufacturing breakthroughs and their translation into commercial products.

National Science Foundation

The National Science Foundation (NSF) supports fundamental advanced manufacturing research, education, and workforce training in its Directorates for Engineering, Computer and Information Science and Engineering, Mathematical and Physical Sciences, and Education and Human Resources. It also promotes advanced manufacturing innovation through a variety of translational research programs, including the Small Business Innovation Research, Small Business Technology Transfer, and Grant Opportunities for Academic Liaison with Industry Programs, and by partnering with industry, states, and other agencies. In fiscal year 2015, the NSF and NIST jointly established and funded MForesight: Alliance for Manufacturing Foresight, a think-and-do tank that harnesses the expertise of the broad U.S.-based manufacturing community to forecast future advanced manufacturing technologies.

The NSF advanced manufacturing investment is primarily through its Cyber-enabled Materials, Manufacturing, and Smart Systems priority area. These programs support fundamental research leading to transformative advances in manufacturing that address size scales from nanometers to kilometers, including process modeling, advanced sensing and control techniques, smart manufacturing using sustainable materials, chemical and biological reactor design and control, and manufacturing processes and enabling technology to support a broad range of industries, with emphases on efficiency, economy, and minimal environmental impact. Advanced manufacturing is also supported through the Engineering Research Centers, Industry/University Cooperative Research Centers, and Advanced Technology Education programs. With an emphasis on two-year colleges, the Advanced Technology Education program focuses on the education of technicians for the high-technology fields that drive our nation's economy.

All NSF programs welcome the submission of proposals to collaborate with Manufacturing Innovation Institutes in cutting-edge research and educational projects. Projects that are currently funded by NSF are also encouraged to request funding supplements to perform collaborate research and/or educational projects with Institutes. It is expected that the incorporation of the resources, expertise, and experience of Institute members will increase the competitiveness of such proposals in merit review.

Small Business Administration

The U.S. Small Business Administration (SBA) was created in 1953 as an independent agency of the Federal Government to aid, counsel, assist, and protect the interests of small business concerns, to preserve free competitive enterprise, and to maintain and strengthen the overall economy of our nation. The SBA recognizes that small business is critical to our economic recovery and strength, to building America's future, and to helping the United States compete in today's global marketplace. Although SBA has grown and evolved since its establishment, the bottom-line mission remains the same. The SBA helps Americans start, build, and grow businesses. Through an extensive network of field offices and partnerships with public and private organizations, SBA delivers its services to people throughout the United States, Puerto Rico, the U.S. Virgin Islands and Guam.

SBA's Office of Investment and Innovation's mission is to stimulate the economy by providing small businesses and entrepreneurs access to capital investment via our Small Business Investment Company (SBIC). The SBIC program is a growth capital program with over $25 billion of assets under management spread across 303 private-public partnerships comprised of private equity, venture capital, and structured lending funds, making it one of the largest fund of funds focused on capitalizing American small businesses.

As of today, roughly 25% of SBIC dollars are invested in manufacturing. In addition, SBA is uniquely suited to play a key part in encouraging the growth of advanced manufacturing in the United States because it plays a central role in realizing the benefits of technological innovation and in the overall growth and health of the U.S. economy. To support this effort, the SBA is proposing the Scale-Up Manufacturing Investment Companies (SUMIC) program to support innovative manufacturing technologies scaling up their first commercial production facilities in the United States. The loan guaranty program would support private funds, operating similarly to the SBIC debt guaranty program, except with a much larger fund and project size necessary to support the needs of manufacturing scale-up efforts.

Investment in small innovative manufacturers will promote the development of cutting edge manufacturing technologies by smoothing the pathway from prototype to production for new processes, tools, and methodologies. By leveraging industry and non-federal co-investment in innovation opportunities that will lead to improved manufacturing capabilities, SUMIC will help to bridge the gap between fundamental technical discoveries in the United States and products manufactured here.

Furthermore, U.S. manufacturers individually are challenged to fund these technology development functions, and small- and medium-size manufacturers struggle with investing in prototyping and scale up of new technologies and potential products. This initiative would help provide the critical mass and knowledge base necessary to address these challenges. Partnerships that bring diverse organizations together to accelerate innovation for advanced manufacturing create a stronger innovation system and link those innovations more directly to domestic production capabilities. This proposal builds on the success of models deployed in other countries.

Appendix B:
Technology Readiness Levels and Manufacturing Readiness Levels

Technology readiness levels (TRLs) and Manufacturing readiness levels (MRLs) are used by the Department of Defense, other Federal agencies, and industry to assess the maturity of new technologies.[51]

TRL 1:	Basic principles observed and reported	MRL 1:	Basic manufacturing implications identified
TRL 2:	Technology concept or application formulated	MRL 2:	Manufacturing concepts Identified
TRL 3:	Experimental and analytical critical function and characteristic proof of concept	MRL 3:	Manufacturing proof of concept developed
TRL 4:	Component or breadboard validation in a laboratory environment	MRL 4:	Capability to produce the technology in a laboratory environment
TRL 5:	Component or breadboard validation in a relevant environment	MRL 5:	Capability to produce prototype components in a production relevant environment
TRL 6:	System or subsystem model or prototype demonstrated in a relevant environment	MRL 6:	Capability to produce a prototype system or subsystem in a production relevant environment
TRL 7:	System prototype demonstration in an operational environment	MRL 7:	Capability to produce systems, subsystems, or components in a production representative environment
TRL 8:	Actual system completed and qualified through test and demonstration	MRL 8:	Pilot line capability demonstrated; ready to begin low rate initial production
TRL 9:	Actual system proven through successful mission operations	MRL 9:	Low rate production demonstrated; capability in place to begin full rate production
		MRL 10:	Full rate production demonstrated and lean production practices in place

---

[51] Manufacturing Readiness Level (MRL) Deskbook: Version 2.4. Prepared by the OSD Manufacturing Technology Program in collaboration with The Joint Service/Industry MRL Working Group. DoD (U.S. Department of Defense). 2015. Washington, DC: U.S. Government Printing Office.

Abbreviations

AMNPO	Advanced Manufacturing National Program Office
ARPA-E	Advanced Research Projects Agency-Energy
BARDA	Biomedical Advanced Research and Development Authority
CBET	Chemical, Bioengineering, Environmental and Transport Systems
CDER	Center for Drug Evaluation and Research
CRISPR	Clustered Regularly-interspaced Short Palindromic Repeats
DARPA	Defense Advanced Research Projects Agency
DMDII	Digital Manufacturing and Design Innovation Institute
DMREF	Designing Materials to Revolutionize and Engineer our Future
DOC	Department of Commerce
DoD	Department of Defense
DOE BETO	Department of Energy Bioenergy Technologies Office
DOE	Department of Energy
DOT	Department of Transportation
EAGER	Early-concept Grants for Exploratory Research
EERE	Energy Efficiency and Renewable Energy
EMN	Energy Materials Network
EPA	Environmental Protection Agency
ERC	Engineering Research Centers
FBI	Federal Bureau of Investigation
FDA	Food and Drug Administration
HHS	Health and Human Services
HPC	High Performance Computing
IACMI	Institute for Advanced Composites Manufacturing Innovation
I/UCRC	Industry/University Cooperative Research Centers
LIFT	Lightweight Innovations for Tomorrow
MDF	Manufacturing Demonstration Facility
MEP	Manufacturing Extension Partnership
MGI	Materials Genome Initiative
MRL	Manufacturing Readiness Level
NASA	National Aeronautics and Space Administration
NCAL	NIST Center for Automotive Lightweighting

NCATS	National Center for Advancing Translational Sciences
NCI	National Cancer Institute
NHLBI	National Heart, Lung and Blood Institute
NIBIB	National Institute of Biomedical Imaging and Bioengineering
NIDDK	National Institute of Diabetes and Digestive and Kidney Diseases
NIFA	National Institute of Food and Agriculture
NIGMS	National Institute of General Medical Sciences
NIH	National Institutes of Health
NIST	National Institute of Standards and Technology
NNI	National Nanotechnology Initiative
NNMI	National Network for Manufacturing Innovation
NNSA	National Nuclear Security Administration
NSF	National Science Foundation
NSTC	National Science and Technology Council
OSTP	Office of Science and Technology Policy
PACT	Production Assistance for Cellular Therapies
PCAST	President's Council of Advisors on Science and Technology
R&D	Research and Development
SAM	Subcommittee on Advanced Manufacturing
SBA	Small Business Administration
SI	International System of Units
STARnet	Semiconductor Technology Advanced Research Network
STEM	Science, Technology, Engineering, and Mathematics
TAACCCT	Trade Adjustment Assistance Community College and Career Training
TRL	Technology Readiness Level
USDA	U.S. Department of Agriculture
VA	Department of Veterans Affairs